Abstracts of the 7th International Con
Durability and Fatigue

FATIGUE 2017

Edited by:
**P. Bailey, F. Berto, E. R. Cawte, P. Roberts,
M. T. Whittaker and J. R. Yates**

*Held at Downing College, University of Cambridge, UK
3rd-5th July 2017*

Engineering Integrity Society

PUBLISHED IN 2017 BY:

Engineering Integrity Society
17 Harrier Close, Cottesmore, Rutland
LE15 7BT, United Kingdom

Tel: +44 (0)1572 811315
Email: info@e-i-s.org.uk
Website: www.e-i-s.org.uk

ISBN 978-0-9544368-2-7

The following publications of the society are available:

FATIGUE 2007 - FATIGUE & DURABILITY ASSESSMENT OF MATERIALS, COMPONENTS & STRUCTURES - Proceedings of the 6th International Conference of the Engineering Integrity Society, Queens' College, Cambridge, UK., March 26-28, 2007, - Editors: M. R. Bache, P. A. Blackmore, E. R. Cawte, P. Roberts and J. R. Yates, ISBN 978-0-9544368-1-0 (2007) 257 pages.

FATIGUE 2003 - FATIGUE & DURABILITY ASSESSMENT OF MATERIALS, COMPONENTS & STRUCTURES - Proceedings of the 5th International Conference of the Engineering Integrity Society, Queens' College, Cambridge, UK., April 7-9, 2003, - Editors: M. R. Bache, P. A. Blackmore, J. Draper, J. H. Edwards, P. Roberts and J. R. Yates, ISBN 0 9544368 0 6 (2003) 566 pages.

Published by the Engineering Integrity Society, 17 Harrier Close, Cottesmore, Rutland, LE15 7BT, U.K. Tel: +44 (0)1572 811315, Email: info@e-i-s.org.uk, www.e-i-s.org.uk

---

FATIGUE 2000 - FATIGUE & DURABILITY ASSESSMENT OF MATERIALS, COMPONENTS & STRUCTURES - Proceedings of the Fourth International Conference of the Engineering Integrity Society, Robinson College, Cambridge, UK., April 10-12, 2000 - Editors: M. R. Bache, P. A. Blackmore, J. Draper, J. H. Edwards, P. Roberts and J. R. Yates, ISBN 1 901537 16 1, 600 pages.

PRODUCT OPTIMISATION FOR INTEGRITY – COMPUTERS: A BOON OR A BURDEN? – Proceedings of the Third International Conference of the Engineering Integrity Society, Sheffield, U.K., April 3-5, 1995 – Editors: E. R. Cawte, J. M. Draper and N. Trigwell, ISBN 0 947817 76 X, 315 pages.

ENGINEERING INTEGRITY IN RAIL TRANSPORT SYSTEMS 1994 – Proceedings of the Third International Conference on Rail Transport Systems – "Rolling Stock Leasing – The Technical Challenges" organised by the Engineering Integrity Society, Birmingham, U.K., July 11-12, 1994 – Editors: J. H. Edwards, A. Jablonski and R. A. Smith, ISBN 0 947817 72 7 (1994), 260 pages.

ENGINEERING INTEGRITY IN RAIL TRANSPORT SYSTEMS 1993 – Proceedings of the Second International Conference on Rail Transport Systems organised by the Engineering Integrity Society, Birmingham, U.K. July 12-13, 1993 – Editor: J. M. Tunna, ISBN 0 947817 65 4 (1993), 412 pages.

ENGINEERING INTEGRITY IN RAIL TRANSPORT SYSTEMS – Proceedings of the First International Conference on Rail Transport Systems organised by the Engineering Integrity Society, Birmingham, U.K., July 13-14, 1992 – Editors: J. M. Tunna and S. J. Hill, ISBN 0 947817 47 6 (1992), 359 pages.

ENGINEERING INTEGRITY THROUGH TESTING – Proceedings of the Second International Conference of the Engineering Integrity Society, Birmingham, U.K., March 20-22, 1990 – Editor: H. G. Morgan, ISBN 0 947817 39 5 (1990), 537 pages.

MEASUREMENT AND FATIGUE – Proceedings of the First International Conference of the Engineering Integrity Society, Bournemouth, U.K., March 17-20, 1986 – Editor: J. M. Tunna, ISBN 0 947817 11 5 (1986), 565 pages.

PREFACE

The 7th International Conference on Durability and Fatigue "FATIGUE 2017" was held at Downing College, University of Cambridge on 3rd - 5th July 2017. This conference series, organised by the Durability & Fatigue Committee of the Engineering Integrity Society, provides a regular interface between leading researchers and practitioners dealing with fatigue issues across the world. The following Keynote presentations were given to introduce a number of the major conference sessions:

"Lifing challenges for gas turbine discs operating at high temperatures"
Steve Williams, Rolls-Royce plc, U.K.

"The early stages of corrosion fatigue: influence of variables on lifetime"
Professor Robert Akid, The University of Manchester, U.K.

"A general probabilistic framework combining experiments and simulations to identify the small crack driving force"
Dr Michael D. Sangid, Purdue University, USA

The success of any conference depends on many factors, not least an enthusiastic and efficient team of organisers in the background. To this end, I must pay tribute to the following for their commitment, support and in the case of the International Scientific Committee, valuable time and effort spent reviewing the papers submitted for the proceedings:

Conference Secretariat:
Mrs Sara Atkin

Conference Organising Committee:
Dr John Yates
Robert Cawte
Paul Roberts
Trevor Margereson
Sara Atkin

Local Technical Committee:
Dr John Yates
Robert Cawte
Paul Roberts
Dr Mark Whittaker
Professor Filippo Berto
Professor Angelo Maligno
Dr Ali Mehmanparast
Dr Peter Bailey
Dr Nicolas Larrosa
Dr Hassan Ghadbeigi
Dr Hayder Ahmad

International Scientific Committee:
Professor Hong Youshi (China)
Professor Filippo Berto (Italy)
Dr Yee Han Tai (UK)
Professor Martin Bache (UK)
Professor Jie Tong (UK)
Professor Luca Susmel (UK)
Professor Ir. Dr. Shahrum Abdullah (Malaysia)

The organisers wish to record their gratitude to all of the presenters, delegates and sponsors for ensuring the success of this meeting and quality of the Conference Proceedings.

Co-sponsoring Institutions:
Institution of Mechanical Engineers

Leading Sponsor:
HBM Prenscia

Conference Sponsors:
Dassault Systèmes Simulia
GOM UK
Instron
Phoenix Calibration & Services Ltd
Severn Thermal Solutions
Siemens
Zwick

For the first time in the history of this meeting the technical papers are provided online.

J. R. Yates
Fatigue 2017 Chairman

Front cover image acknowledgements:

Top row L-R: S. S. K. Singh, S. Abdullah, M. F. M. Yunoh & N. M. N. Abdullah (paper #3.4),
N. C. Barnard, R. J. Lancaster, M. Jones, I. Mitchell & M. R. Bache (paper #1.2)
Middle row L-R: A. L. Dyer, J. P. Jones M. T.Whittaker & R. D. Cutts (paper #4.3),
N.O. Larrosa, C. Evans, J. Carr, U. Tradowsky, N. Read, M. M. Attallah & P. J. Withers (paper #1.4)
Bottom row L-R: C. Steimbreger & M. D. Chapetti (paper #10.1), G. Teixeira, M. Roberts, V. Nascimento, D. Novello & T. Clarke (paper #3.2), T. Borsato, F. Berto, P. Ferro & C. Carollo (paper #9.1)

ENGINEERING INTEGRITY SOCIETY

The Engineering Society offers a unique forum for industrial engineers to exchange ideas and experience. Since its foundation in 1985 the EIS has made an important contribution to engineering science, holding major National and Industrial Conferences, organising and co-ordinating specialist task groups, and presenting technical seminars. Membership is world-wide.

Today's engineers must be able to handle problems which cover a variety of engineering disciplines. The EIS provides a pragmatic and practical arena for engineers who need access to information on a multi-disciplinary basis.

The Engineering Integrity Society publishes bound volumes of Conference Proceedings, and reports of Seminars and Task Groups are published in the Society Journal, which also contains technical articles and papers. All EIS events are open to members and non-members alike.

The Society is a registered charity. Each group has its own organising committee, with representatives on the EIS Council. Financial support is provided by Corporate Members.

Individual membership is open to engineers working in all aspects of research, design, development, analysis and testing. Company membership is welcomed.

For more information on EIS membership and activities:

17 Harrier Close
Cottesmore
Rutland
LE15 7BT
UK

Tel: +44 (0)1572 811315
Email: info@e-i-s.org.uk
Website: www.e-i-s.org.uk

CONTENTS

2.3 Implementation of plasticity model for a steel with mixed cyclic softening and hardening and its application to fatigue assessments
V. Okorokov, T. Comlekci, D. MacKenzie, R. van Rijswick & Y. Gorash

2.4 Effect of the material modelling and the experimental material characterisation on Fatigue Life Estimation within Strain-Based Fatigue Assessment Approaches
M. Hell, R. Wagener, H. Kaufmann & T. Melz

3: VARIABLE AMPLITUDE FATIGUE

3.1 A revised understanding of major load interaction mechanisms in variable-amplitude fatigue
R. Sunder

3.2 Efficient frequency domain fatigue approaches for automotive components
G. Teixeira, M. Roberts, V. Nascimento, D. Novello & T. Clarke

3.3 Fatigue life assessment method based on load spectra considering nonlinear stress behavior and its validation with testing results
C. Donertas, M. Zacharzuck, M. Siktas & B. Ozmen

3.4 The needs to integrate durability assessment using stochastic process for an automobile crankshaft
S. S. K. Singh, S. Abdullah, M. F. M. Yunoh & N. M. N. Abdullah

3.5 Determining probabilistic-based failure of damaging features for fatigue strain loadings
*M. F. M. Yunoh S. Abdullah, M. H. M. Saad, Z. M. Nopiah, M. Z. Nuawi &
S. S. K. Singh*

4: HIGH TEMPERATURE AND THERMOMECHANICAL FATIGUE

4.1 Effect of voids and inclusions on the high temperature localised cyclic behaviour of a next generation power plant material
E. M. O'Hara, N. M. Harrison, B. K. Polomski, R. A. Barrett & S. B. Leen

4.2 Assessment of optical based control methods for thermo-mechanical fatigue
J. P. Jones, S. P. Brookes, M. T. Whittaker, R. J. Lancaster, A. Dyer & S. J. Williams

4.3 Development and validation of a facility to test the thermo-mechanical fatigue behaviour of TiMMCs
A. L. Dyer, J. P. Jones, M. T. Whittaker & R. D. Cutts

4.4 The fatigue behaviour of carburised 316H stainless steel and the application to nuclear plant assessment
S. Flannagan, M. Chevalier, D. Dean, M. Turnock, P. Deem, D. Tsivoulas & A. Wisbey

5: CORROSION FATIGUE

5.1 Keynote Lecture: The early stages of corrosion fatigue: influence of variables on lifetime
R. Akid

5.2 On "chemical notch" concept for corrosion-fatigue life prediction
D. Kujawski, F. Berto & L. Susmel

5.3 A cellular automata finite element analysis (CAFE) modelling approach for predicting the early stages of corrosion under fatigue loading
O. Fatoba, C. Evans, R. Leiva-Garcia, N. Larrosa & R. Akid

5.4 Corrosion fatigue assessment of brazed AISI 304/BNi-2 joints in synthetic exhaust gas condensate
A. Schmiedt, D. Nowak, M. Manka, L. Wojarski, W. Tillmann & F. Walther

5.5 Strain evolution around corrosion pits under fatigue loading using digital image correlation
C. Evans & R. Akid

5.6 Impact of corrosion feature on the low cycle fatigue behaviour of a Ni alloy for disc rotor applications
M. Dowd

6: FATIGUE CRACK GROWTH AND CLOSURE

6.1 Assessment of the effect of overloads on growing fatigue cracks using full-field optical techniques
J. M. Vasco-Olmo & F. A. Diaz

6.2 Study of the stress intensity factor in the bulk of the material with synchrotron diffraction
P. Lopez-Crespo, J. Vazquez-Peralta, C. Simpson, B. Moreno, T. Buslaps & P. J. Withers

6.3 Characterisation of fatigue crack growth behaviour in offshore wind turbine welded structures
A. Mehmanparast, F. Brennan & I. Tavares

7: FATIGUE AT WELDS AND NOTCHES I

7.1 Different approaches for fatigue assessment of butt-welded joints
C. M. Rizzo, F. Berto & T. Welo

7.2 Fatigue assessment of welded joints in large steel structures: a modified nominal stress definition
M. Colussi, F. Berto & G. Meneghetti

7.3 Averaged strain energy density-based synthesis of crack initiation life of notched titanium and steel bars under uniaxial and multiaxial fatigue
A. Campagnolo, G. Meneghetti, F. Berto & K. Tanaka

8: MULTIAXIAL FATIGUE I

8.1 Recent developments on multiaxial fatigue strength of titanium alloys
F. Berto, A. Campagnolo & S. Vantadori

8.2 Accuracy of software Multi_FEAST© in estimating multiaxial fatigue damage in metallic materials
L. Susmel

8.3 Multiaxial low cycle fatigue analysis of notched specimen for type 316L stainless steel under non-proportional loading
T. Morishita, S. Bressan, T. Itoh, P. Gallo & F. Berto

8.4 Effect of interchange in principal stress and strain directions on multiaxial fatigue strength of type 316 stainless steel
Y. Takada, T. Morishita & T. Itoh

9: FATIGUE OF ENGINEERING MATERIALS

9.1 Fatigue properties of solution strengthened ferritic ductile cast irons in heavy section castings
T. Borsato, F. Berto, P. Ferro & C. Carollo

9.2 Contact fatigue characterization of 17NiCrMo6-4 specimens using a twin-disc test bench
A. Terrin & G. Meneghetti

9.3 Overview of copper wire failure mechanisms in rotating machinery
H. Y. Ahmad, D. Bonnieman & D. C. Howard

9.4 Low cycle fatigue behaviour of an ultrafine-grained metastable austenitic CrMnNi steel
M. Droste, M. Fleischer, A. Weidner & H. Biermann

10: FATIGUE AT WELDS AND NOTCHES II

10.1 Significance of undercuts on fatigue strength of butt-welded joints
C. Steimbreger & M. D. Chapetti

10.2 Consideration of weld distortion throughout the identification of fatigue curve parameters using mean stress correction
Y. Gorash, X. Zhou, T. Comlekci, D. MacKenzie & J. Bayyouk

10.3 Calculation of stress concentration factors in offshore welded structures using 3d laser scanning technology and finite element analysis
L. Wang, W. Vesga Rivera, A. Mehmanparast, F. Brennan & A. Kolios

10.4 High-resolution 3D assessment and ACPD crack growth monitoring of weld toe fatigue crack initiation
S. Chaudhuri, J. Crump, P. A. S. Reed, B. G. Mellor, P. Tubby, M. Waitt & A. Wojcik

11: MULTIAXIAL FATIGUE II

11.1 Multiaxial fatigue test and strength using cyclic bending and torsion testing machine
H. Sato, T. Morishita, T. Itoh & M. Sakane

11.2 Numerical study on influence of non-proportional stressing on fretting fatigue life assessment
R. H. Talemi

11.3 The heat energy dissipated in a control volume to correlate the fatigue strength of severely notched and cracked stainless steel specimens
G. Meneghetti & M. Ricotta

12: INFLUENCE OF MANUFACTURING PROCESSES ON FATIGUE

12.1 High cycle fatigue behaviour in ultrasonic-shot-peened β-type Ti alloy and EBSD-assisted fractography of sub-surface fatigue crack initiation
Y. Uematsu, T. Kakiuchi & K. Hattori

12.2 Effect of additive elements on crack initiation and propagation behaviour for low-Ag solders
N. Hiyoshi, M. Yamashita & H. Hokazono

12.3 Fatigue assessment of adhesive wood joints through physical measuring techniques
S. Myslicki, C. Winkler, N. Gelinski, U. Schwarz & F. Walther

13: FATIGUE CRACK GROWTH AND CLOSURE II

13.1 Experimental investigation of thickness effects on fatigue crack closure behaviour in Al7075-T6 alloy
K. Masuda, S. Ishihara & M. Okane

13.2 Through thickness evolution of crack tip plasticity
D. Camas, P. Lopez-Crespo, F. V. Antunes & J. R. Yates

13.3 The Investigation of the Fatigue Crack Growth Rate in Triple Phase Microstructure of a Rotor Steel
A. S. Golezani, M. Mobaraki & R. Samadi

14: FATIGUE OF COMPOSITE MATERIALS AND STRUCTURES

14.1 Fatigue behaviour and mean stress effect of thermoplastic polymers and composites - *Z. Lu, B. Feng & C. Loh*

14.2 How simple is as simple as possible?
P. Heyes

14.3 A progressive damage fatigue model for unidirectional laminated composites based on finite element analysis: theory and practice
M. Hack, D. Carrella-Payan, B. Magneville, T. Naito, Y. Urushiyama, T. Yokozeki, W. Yamazaki, & W. Van Paepegem

14.4 Fatigue behaviour of nylon-clay hybrid nanocomposites
S. J. Zhu, A. Usuki & M. Kato

14.5 Fatigue limits of natural rubber using the carbon black filler materials
M. J. Jweeg, M. Al-Waily, D. C. Howard & H. Y. Ahmad

15: DESIGN AND ASSESSMENT

15.1 Target reliability as an acceptance criterion for fatigue
K. Wright

15.2 Multiaxial fatigue strength prediction criteria – development of validation data sets
J. Papuga, M. Lutovinov & M. Růžička

15.3 Approach to a general bending fatigue strength design method
G. Deng & T. Nakanishi

1: FATIGUE OF ADDITIVE MANUFACTURED MATERIALS

LIFING CHALLENGES FOR GAS TURBINE DISCS OPERATING AT HIGH TEMPERATURES

S. Williams

The need to produce cleaner and more efficient aero gas turbines has led to designs with faster and hotter cores and the development of Ni based powder materials for High Pressure spool compressor and turbine discs. Maintaining component integrity for these novel materials in harsh loading environments has necessitated a step change improvement in Rolls-Royce's understanding of the variables affecting fatigue behaviour and the analytical models used to describe them. In particular, the role of time dependent phenomena such as creep, oxidation and corrosion on both crack nucleation and propagation behaviour has become much more important.

This presentation describes how disc lifing methods have evolved over time, the sometimes unexpected aspects of high temperature material behaviour that need to be characterised and the way that improved understanding is being used to support the declared lives of Critical Parts. In particular it describes the development of non-linear stressing methods to include the effects of creep and plasticity on the mean stress and stress-strain hysteresis loop at a component feature and a correlation approach to lifing that allows all fatigue results from specimens and components to be predicted in a consistent way.

Rolls-Royce plc, PO Box 31, Moor Lane, Derby DE24 8BJ

THE FATIGUE AND CRACK GROWTH CHARACTERISTICS OF ADDITIVE-MANUFACTURED ALLOYS FOR COST EFFICIENT HIGH INTEGRITY AERO-ENGINE COMPONENTS

N.C. Barnard[1]*, R.J. Lancaster[1], M. Jones[2], I. Mitchell[2] & M. R. Bache[1]

A series of fatigue experiments has been performed on the near alpha titanium alloy Ti-6242 produced via a blown-powder additive route. The high cycle fatigue behaviour of additive repair deposits is investigated in the re-machined condition. Damage tolerance characteristics are also reported through the derivation of Stage II crack growth and fatigue crack threshold data. The optimisation of additive materials for fatigue performance is discussed.

[1] Institute of Structural Materials, College of Engineering, Swansea University, Swansea, SA1 8EN
[2] Rolls-Royce plc, PO Box 31, Derby, DE24 8BJ
* Corresponding Author [email: n.c.barnard@swansea.ac.uk; tel: +44 (0)1792 513102]

FATIGUE BEHAVIOR OF TI64 PARTS MADE BY SELECTIVE LASER MELTING PROCESS IN DIFFERENT POST PROCESS CONDITIONS

M. Benedetti[1], M. Cazzolli[1], M. Leoni[2], V. Fontanari[1], M. Bandini[3]

Selective laser melting of Titanium alloys is nowadays a commonly adopted technique for the production of prototypes but suitable also for a real industrial production, especially in the biomedical and aeronautical context. For this kind of applications an extremely important parameter to investigate is the fatigue life of SLM components. Apart from optimal geometry, different aspects should be considered regarding the fatigue life of SLM parts: microstructure, residual stresses, surface roughness and internal porosity.

All these aspects are investigated in the present work in order to have a precise idea about the detrimental effects that each one of them could have on the fatigue resistance. The adopted sample has a specific geometry designed to maximize the occurrence of a critical pore or surface defect in the gauge length. All the fatigue tests were carried out on a Rumul resonant machine at a frequency of 150 Hz. The run out was considered to be 50 millions of cycles to explore damage mechanisms acting in the very high cycle fatigue regime. All the samples after the SLM process are heat-treated (670°C for 5h in protective atmosphere) to relax the residual stress due to the solidification process taking place during the part fabrication. After the thermal treatment a quantitative analysis of the residual stress was made by X-ray diffractometry. The first lot of samples is tested just after the thermal treatment with no other modification. The result of this test is the fatigue life value taken as a benchmark to understand the improvement due to the other processes investigated in this work.

To avoid the presence of internal pores some samples are subject to an HIP (hot isostatic pressing) process. Specifically, samples are heated to 920°C for 2hours at a pressure of 1000 bar. In this case the surface roughness is just slightly affected by the process and can be considered the same as the heat-treated samples. An important effect of the HIP process is the microstructural modification. The isothermal stage at 920°C produces a tempering treatment affecting the final microstructure. To evaluate the differences, metallographic analysis was conducted on both the samples and also the difference in hardness is investigated.

An alternative way to improve the fatigue life is the shot peening process. In the present work, a set of samples was subject to a high intensity shot peening to close surface porosity and create a high compressive stress useful to increase the expected fatigue resistance. The residual stress intensity was measured by X-ray diffractometry. The effect of the surface roughness was also evaluated by electropolishing. The electropolishing step is useful to decrease the surface roughness avoiding any plastic deformation of the surface, usually created adopting a mechanical polishing, that can bring to compressive stress that can misrepresent the final result. The values measured for all the different lots are then compared to select the optimal treatment to be adopted after the SLM process to improve the fatigue life of the produced components.

[1]Department of Industrial Engineering, University of Trento, Trento, Italy
[2]Department of Civil, Environmental and Mechanical Engineering, University of Trento, Trento, Italy
[3]Peen Service Ltd, Bologna, Italy

LINKING PROCESSING DEFECTS AND FATIGUE BEHAVIOUR IN SELECTIVE LASER MELTING OF AlSi10Mg

N.O. Larrosa[1,2]*, C Evans[2], J. Carr[2,3], U. Tradowsky[4], N. Read[5],
M.M. Attallah[5] & P.J. Withers[2,3]

Here we analyse the relationship between the fatigue life of cylindrical AlSi10Mg (CL31 AL) samples fabricated by Selective Laser Melting (SLM) built up parallel to the fatigue axis (VL) and normal to it (HL) in relation to the presence of manufacturing defects (pores, voids, cracks, etc.) and the beneficial effect of post-processing (T6 heat treatment and hot isostatic pressing (HIP)) treatments. X-ray Computed Tomography (CT) is used to characterise the three-dimensional (3D) structure of fatigue samples and to understand the role of defects on the experimental fatigue data. Pancake-shaped pores are observed in the plane of the deposited layers having a 130% higher volume fraction for the VL than the HL orientations. Further, while T6 treatment had relatively little effect, the HIPping reduced the pore fraction by 44% and 65% for VL and HL samples. This increased the 0.2% yield stress. Further the fatigue life at 60% yield stress was significantly greater for the HL oriented samples. Our results suggest that the fatigue life seems to be dominated by the presence of pancake (crack-like) defects perpendicular to the loading direction such that it is better to build samples transverse to the highest fatigue loads. Both T6 heat treatment and HIPping appear to enhance the fatigue behavior of the material regardless of the AM deposition scheme.

*Corresponding author: nicolas.larrosa@bristol.ac.uk
[1]Department of Mechanical Engineering, Solid Mechanics Research Group, University of Bristol, Bristol, BS8 1TR, UK
[2]School of Materials, The University of Manchester, Manchester, M13 9PL, UK
[3]Henry Royce Institute, The University of Manchester, Manchester, M13 9PL, UK.

[4]Photon Laser Engineering, Staakener Str. 53-63, D-13581 Berlin, Germany
[5]School of Metallurgy and Materials, University of Birmingham, Edgbaston, Birmingham, B15 2TT, UK

TEST DEVELOPMENT TO SUPPORT FATIGUE CRACK GROWTH MEASUREMENTS IN FLOW FORMED INCONEL 718

C. Coleman[1]*, M. R. Bache[1] & C. Boettcher[2]

Flow forming technologies are under consideration as cost effective routes for the manufacture of structural axisymmetric components. The final microstructural characteristics and associated mechanical properties are governed by the degree of deformation induced during flow forming and tailored via subsequent heat treatments. Laboratory measurements of the structure-property relationships of flow formed material can be problematic, mainly due to the restrictions imposed on the extraction of conventional specimen geometries since most of the finished tubular or cone structures will contain thin and curved walls. This paper will describe the development of a suitable specimen design and associated test technique for the measurement of fatigue crack growth rates at room and elevated temperatures. Data obtained from flow formed Inconel 718 (IN718) will be compared to specimens of the exact same geometry but machined from conventionally forged IN718 stock. This allowed for validation of the novel flow formed test in addition to an assessment of the damage tolerance of the flow formed variant.

[1]Institute of Structural Materials, College of Engineering, Swansea University, Bay Campus, Fabian Way, Crymlyn Burrows, Swansea SA1 8EN, United Kingdom
[2]Rolls-Royce plc, P.O. Box 31, Derby, DE24 8BJ, United Kingdom

2: MATHEMATICAL AND NUMERICAL MODELLING

A GENERAL PROBABILISTIC FRAMEWORK COMBINING EXPERIMENTS AND SIMULATIONS TO IDENTIFY THE SMALL CRACK DRIVING FORCE

M. D. Sangid[1]*, A. Rovinelli[1], H. Proudhon[2],
R. A. Lebensohn[3], Y. Guilhem[4] & W. Ludwig[5]

Identifying the short crack driving force of polycrystalline engineering alloys is critical to correlate the inherent microstructure variability and the uncertainty in the short crack growth behavior observed during stage I fatigue crack growth. Due to recent experimental advancements, "cycle-by-cycle" data of a short crack propagating through a metastable beta titanium alloy is available via phase and diffraction contrast tomography. To compute the micromechanical fields not available from the experiment, crystal plasticity simulations are performed. Results of the experiment and simulations are combined in a single dataset and sampled utilizing non-local mining technique. Sampled data is analyzed using a machine learning Bayesian Network framework to identify statically relevant correlations between state variables, microstructure features, location of the crack front, and experimentally observed growth rate, in order to postulate a data-driven, non-parametric short crack driving force. Results are presented with a particular focus on the correlation between well-established short driving forces available in literature and the experiments.

[1]Purdue University
[2]MINES ParisTech
[3]Los Alamos National Lab
[4]ENS de Cachan
[5]European Synchrotron Radiation Facility

A SIMPLIFICATION OF A MICROSTRUCTURAL MODEL TO PREDICT FATIGUE LIMITS IN NOTCHES USING FEA

V. Chaves*, C. Madrigal & A. Navarro

Navarro and De los Rios developed a microstructural model based on dislocations to study the growth of short fatigue cracks in a metallic material. The crack growth rate decreases as the crack approaches the grain boundary and accelerates once it has spread into the neighbouring grain. The fatigue limit is defined as the minimum stress that allows the crack to overcome the successive grain boundaries. This method has been applied with success to predict the fatigue limit of some notched geometries. In this paper, an approximation of the model to facilitate its application to potential users from the industry is proposed. It requires the stress gradient ahead of the notch, based on a linear elastic analysis, which can be calculated with a finite element analysis (FEA). This approximation has been applied to circular holes and V-notches and its error with respect to the exact microstructural model is below 12%.

Departamento de Ingenieria Mecanica y Fabricacion. Escuela Tecnica Superior de Ingenieria. Universidad de Sevilla.
Avda. Camino de los Descubrimientos, s/n. 41092. Sevilla.
*E-mail: chavesrv@us.es

IMPLEMENTATION OF PLASTICITY MODEL FOR A STEEL WITH MIXED CYCLIC SOFTENING AND HARDENING AND ITS APPLICATION TO FATIGUE ASSESSMENTS

V. Okorokov[1,a]*, T. Comlekci[1,b,] D. MacKenzie[1,c],
R. van Rijswick[2,d] & Y. Gorash[1,e]

Structural steels used in severe engineering applications show the phenomena of cyclic softening and hardening under cyclic loading conditions. This can be observed as a change of stress at a certain strain level under cyclic conditions when compared to the same stress obtained after simple monotonic tension test. Usually materials show either softening or hardening behaviour when the stresses can only decrease or increase appropriately within the entire strain range. However, some materials, particularly high strength steels, can demonstrate the mixed softening and hardening. In general, softening occurs at the small strains range and hardening occurs at the large strains range. Although cyclic hardening can only take place when stress amplitude exceeds the yield strength of the material, cyclic softening can occur even below the point of plastic yielding. Therefore, for certain strain amplitude the material with elastic response in monotonic condition can be actually subjected to low-cycle fatigue instead of high-cycle. This fact points out the potential issue of fatigue strength overestimation in materials with softening below the monotonic yield strength, which can lead to inaccurate design decisions.

There are a few plasticity theories based on the concept of surface defining memory in the space of plastic strains that can take into account cyclic softening and hardening under cyclic plasticity loading. However, these theories are valid for the description of cyclic behaviour only when the material shows either cyclic softening or hardening when the stress level is above the yield strength. Therefore, it cannot model the mixed softening and hardening quite accurately as well as cyclic softening before the stresses reach the yield point. In order to describe the cyclic behaviour of a material more realistically, a modified nonlinear kinematic hardening plasticity model has been developed. Modifications are mostly based on replacing the constants in standard model on the functions of previously accumulated plastic strain and strain amplitude. The new model can provide a better accuracy of the cyclic material response for increasing or decreasing level test with stabilisation on each strain level. Other phenomena attributed to cyclic loading conditions, such as ratcheting and mean stress relaxation, can also be simulated by the suggested plasticity model.

Numerical simulation of the cyclic material behaviour with proposed plasticity model under uniaxial stress conditions is performed using MAPLE software. More complex problems with multiaxial stress state including fatigue assessments after autofrettage are solved by means of FEA with the use of ANSYS Workbench. The proposed plasticity model together with fatigue equations are incorporated into ANSYS Workbench by the means of User Programmable Features, where user is able to insert custom equations and solving algorithms.

1 – Department of Mechanical and Aerospace Engineering, University of Strathclyde, Glasgow, United Kingdom, G1 1XQ
2 - Weir Minerals, Venlo, the Netherlands, 5928 PH
a* - volodymyr.okorokov@strath.ac.uk, b - tugrul.comlekci@strath.ac.uk,
c - d.mackenzie@strath.ac.uk, d - r.rijswick@weirminerals.com,
e - yevgen.gorash@strath.ac.uk

EFFECT OF THE MATERIAL MODELLING AND THE EXPERIMENTAL MATERIAL CHARACTERISATION ON FATIGUE LIFE ESTIMATION WITHIN STRAIN-BASED FATIGUE ASSESSMENT APPROACHES

M. Hell[1*], R. Wagener[2], H. Kaufmann[2] & T. Melz[2]

Efficient product design processes require fatigue design approaches, which provide methods to exploit the potential of materials regarding lightweight design, durability and sustainability already during draft design. Simultaneously, a reduction of time to market and overall product development expenses raises the need for accelerated numerical fatigue design and experimental testing periods while the demands for accuracy of fatigue life estimation increase continuously. Particularly focusing on the cyclic material behaviour, the application of suitable fatigue design concepts, advanced testing methods and numerical simulations allow assessments of the fatigue behaviour of new materials and the durability of designed components.

In load based linear elastic concepts, e.g. the nominal stress approach, the statistical and mechanical influence of the component geometry, the component size and the influence of boundary layer modifications are implemented using transfer concepts. The transfer concepts, such as the stress gradient approach according to Siebel and Stieler, the approach of the highly stressed volume according to Kuguel or the weakest link approach according to Weibull, differ substantially in regard to the definition of fatigue notch effects and size effects. Therefore it is not possible to quantitatively assess singular effects separately.

Strain based fatigue design concepts involve an elasto-plastic material model and are based on the assumption, that the material behaviour for an infinitesimal small material volume may be derived from strain controlled tests with macroscopic strain measurements. As recent investigations indicate, the assumption of size effects also applies to strain based concepts. Furthermore a separation between different size effects is possible. Statistical size effects may be implemented using a modified weakest link approach,whereas notch support effects are assessed using elasto-plastic finite element analyses on the basis of the cyclic stress-strain behaviour. In contrast to load based concepts, which assume linear elastic material behaviour, the correlation parameters of the transfer concepts, as for example highly stressed volumes or areas, depend on the load-time history and the cyclic stress-strain behaviour, especially when the strain exceeds the yield limit locally. The cyclic stress-strain behaviour includes cyclic hardening or softening and, depending on the load ratio and the boundary conditions of the cyclic deformation, also cyclic ratchetting and mean stress relaxation. For numerical modelling of the cyclic material behaviour accurately, numerous material models have been proposed. The presented work aims at an evaluation of the impact of different selected material models on the numerical assessment of the local stress strain state and the evaluation of size effects, taking also into account the validation of the material models with experimental data. As a comparison between the numerical assessed fatigue life and the experimental results show, using an appropriate material model in connection with an implementation of size effects may reduce the numerical as well as the experimental effort.

[*1]Chair for System Reliability and Machine Acoustics, Technische Universität Darmstadt, Magdalenenstrasse 47, 64289 Darmstadt, Germany, email-adress: hell@szm.tu-darmstadt.de
[2]Fraunhofer Institute for Structural Durability and System Reliability

3: VARIABLE AMPLITUDE FATIGUE

A REVISED UNDERSTANDING OF MAJOR LOAD INTERACTION MECHANISMS IN VARIABLE-AMPLITUDE FATIGUE

R. Sunder

Major load interaction mechanisms said to influence variable-amplitude fatigue crack growth include crack closure, crack-tip residual stress and blunting/re-sharpening. However, conventional analytical models used to estimate residual fatigue life are based only on one of the first two mechanisms. It would follow that realistic estimates are possible either because other interaction mechanisms are insignificant, or, because their effect is somehow 'fudged' by appropriate corrections to the effect of the particular mechanism that was modeled.

Targeted research over the years involving mechanism isolation through specially designed experiments followed by quantitative electron fractography has provided an improved understanding of how exactly different load interaction mechanisms operate. It provides the basis to fine-tune conventional methodology by accounting for important details neglected by the more simplistic procedures and equations in place.

A major breakthrough in understanding the residual stress effect on metal fatigue was provided by the discovery that the effect practically vanishes in high vacuum [1]. More experiments showed that the effect is restricted to the near-threshold regime, thereby underscoring a major shortcoming of prevailing residual stress models that equate it with the stress ratio effect. Further research involved the development of a simple analytical model to compute near-tip residual stress as a function of applied load history, e.g., application of overloads. This, combined with an experimental procedure to estimate ΔK_{th} after periodic overloads revealed a close connection between computed near-tip residual stress and ΔK_{th} [2]. It follows, that modeling the residual stress mechanism in atmospheric variable-amplitude fatigue demands cycle-by-cycle computation of near-tip residual stress, followed by suitable correction of ΔK_{th} for the following rising load half-cycle. Conventional models resort to a less subtle 'retardation coefficient' because they did not have the benefit of our new understanding of the root cause.

Crack-tip blunting after a tensile overload was originally thought to be a reason for retarded crack growth. However, blunting also keeps the crack open by not allowing wake contact – an effect confirmed by delayed retardation after a tensile overload [3]. This raises the question of post overload closure transient associated with crack-tip blunting. A recent fractographic study of crack growth under specially programmed load sequences suggests that closure can recover rapidly under periodic overloads, indeed, much more rapidly than in the case of single overload. This discovery was possible thanks to isolation of ΔK_{th} sensitivity to local residual stress, a detail that deserves consideration for improved modeling of the closure effect.

The effect of fatigue crack closure is typically modeled through estimation of a certain Kop below which, the fatigue crack is deemed closed. However, in reality, closure is a process, not just a point on the load cycle. The difference would be telling on those smaller load cycles that are applied in the vicinity of Kop, where near-tip stress-strain response is not exactly the same as it would be for a fully open crack. Real load spectra involve a vast majority of small cycles and this is a good reason to pay more attention to actual crack-tip response under conditions of 'partial crack closure'. Work is in progress to unravel closure as a process.

BiSS (P) Ltd, 497E 14th Cross, 4th Phase, Peenya Industrial Area, Bangalore 560058, India, E-mail: rs@biss.in, Mob: +91-9880-432-322

EFFICIENT FREQUENCY DOMAIN FATIGUE APPROACHES FOR AUTOMOTIVE COMPONENTS

G. Teixeira[1], M. Roberts[1] , V. Nascimento[2],
D. Novello[2] & T. Clarke[3]

In random vibration the loads are best described in terms of the PSD (Power Spectrum Density) of the variable exciting the system, whereas if the loads are deterministic the time domain based approaches are usually preferred. As the Finite Element (FE) Method has become the most popular tool for materials modelling, in structural dynamics the modal and harmonic analyses are the starting point for studying system properties and deriving transfer functions (FRF), which are the key to evaluating the spectral moments of stress or strain PSDs and building probability density functions (PDF) to evaluate damage. Recent researches have endeavoured to address residual and mean stresses in the frequency domain and cover the gaps in Dirlik's approach. The present paper is mostly dedicated to provide further details on a particular method known as the Tovo and Benasciutti Method (TB). A case study is presented where a brake chamber has been analyzed and the numerical results compared to laboratory and off-road field tests.

[1]Dassault Systemes SIMULIA, Peel Street, Sheffield, UK
[2]Master Sistemas Automotivos , Rua Atilio Andreazza 3520, Caxias, RS, Brazil
[3]UFRGS, Av. Bento Goncalves, 9500, CADETEC - Porto Alegre, RS, Brazil

FATIGUE LIFE ASSESSMENT METHOD BASED ON LOAD SPECTRA CONSIDERING NONLINEAR STRESS BEHAVIOUR AND ITS VALIDATION WITH TESTING RESULTS

C. Donertas[1], M. Zacharzuck[2], M. Siktas[1] & B. Ozmen[1]

The aim of the study is to improve the quality of predictions regarding the structural durability at an early stage of development of a steel leaf spring. It is important to consider all significant nonlinear factors to analyze the fatigue life. So far, mostly linear effects are taken into account in the calculation of steel leaf spring damages. This study deals with the inclusion of nonlinear factors in the calculation of fatigue life and presents a good correlation with testing results. At the beginning, static simulations are run considering significant nonlinearities to get the stress values. These stress results are validated with strain gauge measurements. Then the fatigue life is calculated with MATLAB-program that includes nonlinearities. These fatigue life results are also compared with a commercial software (FEMFAT) and testing results. The study shows that the consideration of nonlinearities lead to an improvement of the calculated results of steel leaf springs to a certain extent.

[1]Mercedes-Benz Turk, Mercedes Bulvari No.17 Orhangazi Mah. Esenyurt, Istanbul 34519, Turkey
[2]University of Stuttgart, Pfaffenwaldring 27, 70569 Stuttgart, Germany

THE NEED TO INTEGRATE DURABILITY ASSESSMENT USING STOCHASTIC PROCESSES FOR AN AUTOMOBILE CRANKSHAFT

S. S. K. Singh[1,2], S. Abdullah[1,2], M. F. M. Yunoh[1,2] & N. M. N. Abdullah[3]

This paper presents the establishment of an integrated approach for durability assessment in fatigue life prognosis under random loading for an automobile crankshaft. The integrated approach proposes the use of an embedded Markov process by incorporating loading data samples to artificially generate a loading history to predict the fatigue life cycle assessment of a crankshaft. The Markov process continuously updated the loading history data in order to reduce the credible intervals between each data point for fatigue life assessment through the linear fatigue damage accumulation rule. The accuracy of the integrated approach was validated through its statistical correlation properties and finite element analysis based on the design and material of the component. The integrated approach corresponded well by presenting an accuracy of 95% when compared to the fatigue data of the material based on the actual sampling data obtained from an automobile industry. Therefore, the use of a non-deterministic method provides a basis for the development of a safer and more reliable durability assessment between the theory and practice of the deterministic method.

[1]Department of Mechanical & Materials Engineering, Faculty of Engineering & Built Environment, Universiti Kebangsaan Malaysia, 43600 Bangi Selangor Malaysia
[2]Centre for Automotive Research, Faculty of Engineering & Built Environment, Universiti Kebangsaan Malaysia, 43600 Bangi Selangor Malaysia
[3]Department of Mechanical Engineering, Universiti Malaysia Pahang Lebuhraya Tun Razak 26300 Gambang, Pahang, Malaysia
Corresponding author: S.Abdullah; email: shahrum@ukm.edu.my
Contact No:+603 89118411; Fax: +603 89259659

DETERMINING PROBABILISTIC-BASED FAILURE OF DAMAGING FEATURES FOR FATIGUE STRAIN LOADINGS

M. F. M. Yunoh[1,2], S. Abdullah[1,2]*, M. H. M. Saad[1], Z. M. Nopiah[1], M. Z. Nuawi[1] & S. S. K. Singh[3]

This paper presents the behaviour of fatigue damage extraction in fatigue strain histories of automotive components using the probabilistic approach. This is a consideration for the evaluation of fatigue damage extraction in automotive components under service loading that is vital in a reliability analysis. For the purpose of research work, two strain signals data are collected from a car coil spring during a road test. The fatigue strain signals are then extracted using the wavelet transform in order to extract the high amplitude segments that contribute to the fatigue damage. At this stage, the low amplitude segments are removed because of their minimal contribution to the fatigue damage. The fatigue damage based on all extracted segments is calculated using some significant strain-life models. Subsequently, the statistics-based Weibull distribution is applied to evaluate the fatigue damage extraction. It has been found that about 70% of the probability of failure occurs in the 1.0 x 10-5 to 1.0 x 10-4 damage range for both signals, while 90% of the probability of failure occurs in the 1.0 x 10-4 to 1.0 x 10-3 damage range. Lastly, it is suggested that the fatigue damage can be determined by the Weibull distribution analysis.

[1]*Department of Mechanical & Materials Engineering, Faculty of Engineering & Built Environment, Universiti Kebangsaan Malaysia,43600 UKM, Bangi, Selangor, Malaysia.
[2]Centre for Automotive Research (CAR), Faculty of Engineering & Built Environment, Universiti Kebangsaan Malaysia,43600 UKM, Bangi, Selangor, Malaysia.
[3]Department of Mechanical Engineering, Politeknik Ungku Omar, Jalan Raja Musa Mahadi, 31400, Ipoh, Perak, Malaysia.

4. HIGH TEMPERATURE AND THERMOMECHANICAL FATIGUE

EFFECT OF VOIDS AND INCLUSIONS ON THE HIGH TEMPERATURE LOCALISED CYCLIC BEHAVIOUR OF A NEXT GENERATION POWER PLANT MATERIAL

E. M. O'Hara[1,6], N. M. Harrison[2,6], B. K. Polomski[3],
R. A. Barrett[4,6] & S. B. Leen[5,6]

The key step for next generation power plants is the development of advanced materials capable of achieving high flexibility and efficiency at increased steam temperatures and pressures. Such operating conditions will cause increased fatigue and creep degradation of plant components, where a key limitation of operating under such conditions is the capability of the current generation of materials. Consequently, multi-scale characterisation via experimental and computational methods is necessary to both characterise and predict next generation material behaviour under flexible plant operating conditions at increased temperatures. MarBN is a new precipitate-strengthened 9-12Cr martensitic steel, with improved strength and microstructure stabilisation under long-term loading via increased tungsten solute strengthening and boron enriched grain boundary precipitates.

A combined work program of experimental testing, microstructural analysis and computational modelling on a cast MarBN material is presented with comparisons to current state-of-the-art material, P91 steel. An experimental program of high temperature low cycle fatigue (HTLCF) tests is conducted at 600 °C, with microstructural analysis to (i) identify the key mechanisms of deformation and (ii) characterise the evolution of the hierarchical microstructure. A material model, which incorporates damage modelling and life prediction, is calibrated and validated across a range of loading conditions at high temperature, and allows prediction of the constitutive behaviour of the cast MarBN material with good correlation against measured data.

[1]E.OHara2@nuigalway.ie, Ph.D. Researcher, Mechanical Engineering, NUI Galway, H91 HX31, Ireland

[2]Noel.Harrison@nuigalway.ie, Lecturer, Mechanical Engineering, NUI Galway, H91 HX31, Ireland
[3]Bartosz-k.Polomski@power.alstom.com, Materials Engineer, GE Power, Gas Power Systems, Newbold Road, Rugby, Warwickshire, CV21 2NH, United Kingdom
[4]Richard.Barrett@nuigalway.ie, Post-Doctoral Researcher, Mechanical Engineering, NUI Galway,
H91 HX31, Ireland
[5]Sean.Leen@nuigalway.ie, Professor of Mechanical Engineering, NUI Galway, H91 HX31, Ireland
[6]Ryan Institute for Environmental, Marine and Energy Research, NUI Galway, H91 HX31, Ireland

ASSESSMENT OF OPTICAL BASED CONTROL METHODS FOR THERMO-MECHANICAL FATIGUE

J. P. Jones[1], S. P. Brookes[2], M. T. Whittaker[1], R. J. Lancaster[1], A. Dyer[1] & S. J. Williams[3]

Thermo-mechanical fatigue (TMF) testing has proven to be a difficult mechanical test to perform due to the inherent complexity in accurately controlling dynamic temperatures. Whilst many researchers favour thermocouple controlled techniques, the current paper argues for the use of optical based methods which can provide direct temperature measurement from the critical volume of material, provided issues regarding variation in emissivity can be overcome. It is shown that under general TMF setups the use of thermocouples is limited by their positioning and susceptibility to temperature spikes, along with temperature shielding during thermal profiling, thus rendering them unsuitable for the application.

[1]Institute of Structural Materials, Swansea University Bay Campus, SA1 8EN
[2]Rolls-Royce, Mechanical Test Operations Centre, GmbH, Germany
[3]Rolls-Royce plc, Derby, DE24 8BJ

DEVELOPMENT AND VALIDATION OF A FACILITY TO TEST THE THERMO-MECHANICAL FATIGUE BEHAVIOUR OF TiMMCS

A. L. Dyer[1], J. P. Jones[1], M. T. Whittaker[1] & R. D. Cutts[2]

Titanium Metal Matrix Composites (TiMMCs) have a significant role in the future of gas turbine development, representing sizeable potential weight savings within applications such as compressor discs. Continuously reinforcing a titanium matrix with Silicon Carbide (SiC) fibres produces a material with the strength, stiffness and creep resistance of SiC, the damage tolerance of titanium and sizeable weight savings compared to monolithic Ti.

However, thermal residual stresses are known to be present in the matrix and the composite reinforcement. This is due to differing values of the thermal expansion coefficient and the manufacturing process. These residual stresses influence the resulting mechanical performance of the material. Isothermal tests have shown that room temperature lives are reduced compared to those at higher temperatures due to the combination of loading stress and residual stress.

The material characteristics have been investigated under thermomechanical fatigue (TMF) loading conditions. TMF tests differ significantly from isothermal tests as both thermal and mechanical loads are controlled individually depending on a desired phasing, deemed the phase angle.

TiMMC specimens were tested under an 80-300°C thermal cycle over a range of stresses. 'In phase' cycles (IP, 0°), where the peak load coincides with the peak temperature, and 'out-of-phase' (OP, 180°) cycles, where the peak load coincides with the minimum temperature, were employed.

In this paper stress controlled strain monitored TMF is considered, allowing for the onset of plastic deformation and creep to be seen through the evolution of strain for both IP and OP conditions. This testing technique allows the fundamental behaviour of the material to be investigated under complex loading conditions and future life prediction methodology to be improved.

[1]Institute of Structural Materials, Bay Campus, Swansea University, Crymlyn Burrows, Swansea SA1 8EN, UK
[2]Rolls-Royce plc, PO Box 31, Derby, DE24 8BJ, UK

THE FATIGUE BEHAVIOUR OF CARBURISED 316H STAINLESS STEEL AND THE APPLICATION TO NUCLEAR PLANT ASSESSMENT

A. Wisbey[1], M. Chevalier[2], D. Dean[2], P. Deem[3], M. Lynch[1], J. Eaton-McKay[1], M. Turnock[3]

The UK Advanced Gas Cooled Reactor (AGR) fleet of stations are unusual in operating in both the creep and fatigue regime of the main structural materials, of which 316H austenitic stainless steel is a significant proportion. This has required the development of a suite of plant life assessment tools, known as the R5 code, to account for these damage mechanisms. This code continues to develop and recently this has begun to examine the impact of a hardened surface layer (via carburisation) on the 316H steel. In this paper the effect of fatigue cycling has been considered.

316H steel test samples have been pre-conditioned by exposure to pressurised CO_2 (AGR coolant) at elevated temperature, producing a reasonably uniform carburised surface layer. These samples have been subjected to strain controlled fatigue testing at 550°C, with comparative tests on as received material also performed. These tests show that at high strain ranges there may be a reduction in life with surface carburisation but this surface layer may be beneficial at low strain ranges. Fractography suggests that the carburised layer promotes multiple crack initiation sites, rather than the limited number associated with conventional parent material. To enable the cyclic deformation behaviour of the carburised material to be deconvoluted from the bulk parent a number of hollow bar test pieces were produced and tested in the same manner as the standard test pieces. The carburised layer showed a significant increase in the cyclic strength compared with the bulk parent.

Potential routes to incorporate the fatigue behaviour into the R5 assessment code have been considered, along with validation procedures.

[1]High Temperature Materials, Amec Foster Wheeler, Walton House, Birchwood Park, Warrington, Cheshire. WA3 6GA, UK.
[2]EDF Energy Nuclear Generation Ltd, Barnett Way, Barnwood, Gloucester, GL4 3RS, UK.

[3]Applied Chemistry & Materials, Amec Foster Wheeler, Building 601, Birchwood Park, Warrington, Cheshire. WA3 6GN
Corresponding author – Dr.Andrew Wisbey,
email – andrew.wisbey@amecfw.com

5. CORROSION FATIGUE

THE EARLY STAGES OF CORROSION FATIGUE: INFLUENCE OF VARIABLES ON LIFETIME

R. Akid

Cyclic mechanical fatigue and electrochemical corrosion combine to produce the synergistic phenomenon described as 'Corrosion Fatigue'. The combination of stress and corrosion is more damaging than that of the sum of the individual mechanisms acting separately and therefore any attempt to investigate and predict this mechanism requires consideration of this synergism.

It is well recognised that the fatigue lifetime of components having a high integrity surface finish is dominated by the development and propagation of short cracks, with the so-called 'initiation' stage accounting for up to 70-80% of the fatigue lifetime under 'inert' test conditions. It is therefore not surprising that any factor that affects this stage, most notably corrosion, will, in turn, affect this stage and ultimately the fatigue life of the component.

The intention of this presentation is to overview the effects of the various mechanical and electrochemical variables on the 'initiation' stage and the important transition of a pit to a crack for corrosion fatigue loading conditions. In particular examples of the influence of microstructural condition, loading frequency, chemical composition of environment, electrochemical potential and uniaxial and biaxial loading conditions will be presented. In addition comments on the approaches to modelling CF from smooth surfaces will be given.

Corrosion & Protection Centre, School of Materials, The University of Manchester, Sackville Street, Manchester M1 3BB

ON "CHEMICAL NOTCH" CONCEPT FOR CORROSION-FATIGUE LIFE PREDICTION

D. Kujawski*[1], F. Berto[2,3] & L. Susmel[4]

Despite the existence of complex elastic-plastic finite element analysis (FEA) software, simplified linear-elastic FEA calculations are widely used in industry for fatigue life predictions because they have a number of advantages over more accurate but complex and time consuming elastic-plastic analysis. The main advantages of the linear-elastic analysis are: (1) speed, (2) easy to use, (3) scalability, and (4) solutions don't require time consuming iterations. It is well recognized that environment plays a significant role on the failure of cyclically loaded components/structures. Existing experimental data indicates that fatigue limit σ_{FL} of a smooth specimen is significantly lower and fatigue life is much shorter in corrosive environment that in more inert environment such as a dry air or vacuum. This paper proposes a novel concept of "chemical notch" and its application to corrosion-fatigue life prediction using stress-based approach. A corrosion fatigue factor k_{corr} is defined as the ratio of the fully-reversed stress amplitude in air, $(\sigma_a)_{air}$, over that in corrosive environment, $(\sigma_a)_{corr}$, for a given fatigue life in terms of a number of cycles to failure, N_f, i.e. $k_{corr} = (\sigma_a)_{air}/(\sigma_a)_{corr}$ at the same N_f. The corrosion fatigue factor resembles the widely used fatigue notch factor k_f. The proposed strategy for corrosion-fatigue life prediction requires the S-N curve in air and the corresponding k_{corr} factor. Preliminary experimental data on 7075-T651 Al tested in laboratory air and 3.5% of NaCl solution and Ti-6Al-4V alloy tested in air and methanol are used to illustrate and validate the proposed method. A fairly good agreement is demonstrated in terms of the correlation among air and corrosion-fatigue life data.

*Corresponding author email: daniel.kujawski@wmich.edu
[1]Western Michigan University, Mechanical and Aerospace Eng., Kalamazoo MI, USA
[2]University of Padua, 36100 Vicenza, Italy
[3]Department of Engineering Design and Materials, NTNU, Norway

[4]University of Sheffield, Department of Civil and Structural Engineering, Sheffield, United Kingdom

A CELLULAR AUTOMATA FINITE ELEMENT ANALYSIS (CAFE) MODELLING APPROACH FOR PREDICTING THE EARLY STAGES OF CORROSION UNDER FATIGUE LOADING

O. Fatoba[1], C. Evans[2], R. Leiva-Garcia[1], N. Larrosa[3] & R. Akid[1]

Modelling localised corrosion presents significant challenges, especially where several complex, non-linear, processes operate at different length scales. Additionally, the challenge to model these complexities is further increased when there is simultaneous interaction between corrosion and stress, notably during corrosion fatigue, where the latter can lead to increased electrochemical activity. The CAFE approach described here offers a new methodology to predict the progress of the time dependent corrosion process occurring under mechanical loading. This in turn opens the opportunity to predict the early stages of corrosion fatigue including the critical pit-to-crack transition stage.

Cellular automata (CA) are discrete computational systems in which the evolution of the state of each cell in the modelling space is determined by the current state of the cell and that of its neighbourhood cells. In this work, the cumulative mechano-electrochemical damage process is decoupled into corrosion and mechanical components, which will then be modelled using cellular automata (CA) and finite element method (FE) respectively. These two systems are then coupled in such a way that there is information flow between them. This modelling approach is validated using experimental data obtained from pitting data and strain maps obtained from Digital Image Correlation (DIC) experiments. DIC is an optical method that employs tracking and image registration techniques for accurate 2D and 3D measurements of changes in images. Therefore, when this system is coupled with mechanical testing, i.e. fatigue tests, strain maps around geometrical defects (pits) on the specimens under fatigue may be obtained. The DIC results can be used to determine the threshold strain value for the pit-crack transition.

The results of this work will increase the understanding of the interaction between corrosion and mechanical loading during the early stages of corrosion fatigue. Specifically, the evolution with time of damage mechanisms, the morphology of localised defects and stress distribution and the dependencies between them will be studied.

[1]Corrosion and Protection Centre, School of Materials, University of Manchester, Sackville Street, Manchester M13 9PL, UK.
[2]BAE Systems Submarines, Barrow, Cumbria, LA14 1AF, UK
[3]Faculty of Engineering. University of Bristol, Clifton Bristol, BS8 1TR (UK)

CORROSION FATIGUE ASSESSMENT OF BRAZED AISI 304/BNI-2 JOINTS IN SYNTHETIC EXHAUST GAS CONDENSATE

A. Schmiedt[1], D. Nowak[1], M. Manka[2], L. Wojarski[2], W. Tillmann[2] & F. Walther[1]

Brazing is considered as an economic joining technology that is even applicable for not weldable material combinations and therefore established in a wide range of industrial applications, such as automotive parts. During the operation of e.g. exhaust gas heat exchangers, the fatigue loading of brazed components is superimposed by corrosive attack due to aggressive exhaust gases.

In the present study the influence of corrosion on the microstructure and the depending mechanical properties under cyclic loading of austenitic stainless steel AISI 304 brazed at 1,050 °C with nickel-based filler metal BNi-2 is investigated. The corrosion fatigue behavior was characterized in synthetic test condensates with composition according to VDA test sheet 230-214, which is established for condensate corrosion in exhaust gas-carrying components. Brazed joints were cyclically tested in corrosive environments using a corrosion cell as well as in air after pre-corrosion due to ageing durations of up to six weeks to assess the influence of superimposed and successive corrosion fatigue loading. Therefore, cyclic stepwise load increase tests complemented with plastic strain, electrical resistance and electrochemical measurements were applied to estimate the fatigue as well as corrosion fatigue properties. The results have been validated in constant amplitude tests until $2 \cdot 10^6$ cycles. Corrosion- and deformation-induced microstructural changes of base materials and joining zones were evaluated using light and scanning electron microscopy.

A significant reduction of fatigue strength at $2 \cdot 10^6$ cycles down to 43% was determined for superimposed and down to 22% for successive corrosion fatigue loading. Local plastic strain measurements using optimized extensometers as well as deformation-induced changes in electrical resistance and electrochemical measurements have proven to be appropriate for a precise corrosion fatigue assessment of brazed AISI 304/ BNi-2 joints.

[1]TU Dortmund University, Department of Materials Test Engineering (WPT), Baroper Str. 303, D-44227 Dortmund, Germany, anke.schmiedt@tu-dortmund.de
[2]TU Dortmund University, Chair of Materials Engineering (LWT), Leonhard-Euler-Str. 2, D-44227 Dortmund, Germany

STRAIN EVOLUTION AROUND CORROSION PITS UNDER FATIGUE LOADING USING DIGITAL IMAGE CORRELATION

C. Evans[1] & R. Akid[2]

Localised corrosion plays a dominant role in the failure of offshore riser systems through the mechanism of corrosion fatigue. The API-5L X65 pipeline steel that is often used to construct riser systems shows good fatigue resistance when tested in air. However, under the synergistic effect of a corrosive environment and cyclic stress, this fatigue resistance is greatly diminished.

Localised corrosion, in the form of pitting, occurs in the early stages of corrosion fatigue. The corrosion pits that initiate and grow in the walls of pipelines can act as precursors to cracking due to a stress concentration around the pits. The pit-crack transition stage of the damage process occurs once the corrosion pit is of a size/geometry where it creates a sufficient stress concentration to allow generation of localised strain, which in turn leads to the initiation of a crack. Therefore, it is of interest to determine the effect of pit geometry on the fatigue resistance of the pipeline steel.

In this study, single corrosion pits were created using a micro-electrochemical cell, which allowed geometrical control to be established. This method enables the effect of a single corrosion pit on fatigue lifetime to be studied. Artificial pits were then created in the surface of both tensile and fatigue specimens manufactured from API-5L X65 (0.08 wt% C).

In-situ Digital Image Correlation (DIC) was performed on pre-pitted fatigue specimens to examine the evolution of strain around artificial corrosion pits across a range of pit-depths. The aim of the study was to establish whether there is a critical strain concentration at which a crack initiates from a corrosion pit, and whether this critical strain value is dependent on pit geometry. Following this, two pits were generated on fatigue specimens using different pit-depth and separation distance combinations. DIC was then used to investigate the interaction between neighbouring pits, and its effect on surface strain evolution during crack initiation and subsequent crack coalescence.

[1]BAE Systems Submarines, Barrow-in-Furness, Cumbria, LA14 1AF (Formerly Corrosion Protection Centre, School of Materials, University of Manchester)
[2]Corrosion Protection Centre, School of Materials, The Mill, Sackville Street, Manchester, M1 3AL

IMPACT OF CORROSION FEATURES ON THE LOW CYCLE FATIGUE BEHAVIOUR OF A NI-ALLOY FOR DISC ROTOR APPLICATIONS

M. Dowd[1], K. M. Perkins[2] and D. J. Child[3]

Currently there is doubt surrounding the suitability of chemically induced stress independent pre-conditioning as thick oxide scales and pit coalescence can lead to net section loss and typically a lack of grain boundary sulphide attack seen in components that experience stress. This research uses micro-notching prior to low salt flux corrosion to precipitate grain boundary sulphide particles within channel-like features akin to stress assisted corrosion morphologies. On fatigue testing, a critical stress was identified where a change of mechanism was observed. The grain boundary oxide formed in the wake of sulphide particles fractures around segments of grains leading to a metal loss that contributes to a significant reduction in fatigue properties. Surface mechanical properties are also affected as fresh sulphide particle formation is expected to drive the recrystallisation of local microstructure through reacting with scale forming and gamma-prime stabilising elements.

[1]Institute of Structural Materials, Swansea University, Bay Campus, Swansea, SA1 8EN, UK. (Corresponding author: M.Dowd@swansea.ac.uk)
[2]College of Engineering, Swansea University, Bay Campus, Swansea, SA1 8EN, UK.
[3]Rolls-Royce plc., P.O. Box 31, Derby, DE24 8BJ, UK.

6. FATIGUE CRACK GROWTH AND CLOSURE

ASSESSMENT OF THE EFFECT OF OVERLOADS ON GROWING FATIGUE CRACKS USING FULL-FIELD OPTICAL TECHNIQUES

J. M. Vasco-Olmo[1] & F. A. Díaz[1]

In this work, the retardation effect induced on growing fatigue cracks under different overloads is evaluated from the analysis of stress intensity factors (SIFs). Full-field optical techniques, namely transmission photoelasticity and digital image correlation (DIC), are used to calculate stress intensity factors from the analysis of stress and displacement crack tip fields. Thus, stress crack tip fields are analysed in the case of photoelasticity, while displacement fields are analysed by DIC. A novel mathematical model (CJP model), that incorporates the effects of plasticity during fatigue crack growth, is implemented to characterise stress/displacement fields at the vicinity of the crack tip. It postulates that the plastic enclave that exists around the tip of a fatigue crack and along its flanks will shield the crack from the full influence of the elastic stress field. Fatigue cracks were grown in compact tension specimens made from polycarbonate (for photoelasticity) and 2024-T3 aluminium alloy (for DIC). In addition, different overload levels were applied during the crack propagation. The retardation effect induced by overloads was quantified from the estimation of the crack opening loads. Moreover, a compliance based method is employed to compare and validate those results obtained by DIC. Results show a good level of agreement, highlighting the potential of the evaluated optical techniques to study fracture mechanics problems. In addition, the adopted mathematical model demonstrates to be a valuable tool in the study of plasticity-induced crack shielding. Results presented in the current work intent to contribute to a better understanding of the shielding effects of crack tip plasticity during fatigue crack growth.

[1]Departamento de Ingeniería Mecánica y Minera, Escuela Politécnica Superior de Jaén, University of Jaén, Campus Las Lagunillas, Jaén 23071, Spain.
Corresponding author: Dr. José Manuel Vasco Olmo, e-mail: jvasco@ujaen.es
Co-author: Dr. Francisco Alberto Díaz Garrido, e-mail: fdiaz@ujaen.es

STUDY OF THE STRESS INTENSITY FACTOR IN THE BULK OF THE MATERIAL WITH SYNCHROTRON DIFFRACTION

P. Lopez-Crespo[1] , J. Vazquez-Peralta[1], C. Simpson[2], B. Moreno[1],
T. Buslaps[3] & P. J. Withers[2]

In this work we present the results of a hybrid experimental and analytical approach for estimating the stress intensity factor. It uses the elastic strains within the bulk obtained by synchrotron X-ray diffraction data. The stress intensity factor is calculated using a multi-point over-deterministic method where the number of experimental data points is higher than the number of unknowns describing the elastic field surrounding the crack-tip. The tool is tested on X-ray strain measurements collected on a bainitic steel. In contrast to surface techniques the approach provides insights into the crack tip mechanics deep within the sample.

[1]Department of Civil and Materials Engineering, University of Malaga, C/Dr Ortiz Ramos, s/n, 29071 Malaga, Spain
[2]School of Materials, University of Manchester, Grosvenor Street, Manchester M1 7HS, UK
[3]ESRF, 6 rue J Horowitz, 38000 Grenoble, France

CHARACTERISATION OF FATIGUE CRACK GROWTH BEHAVIOUR IN OFFSHORE WIND TURBINE WELDED STRUCTURES

A. Mehmanparast[1], F. Brennan[1] & I. Tavares[2]

An important issue to be understood in the life assessment of offshore renewable energy wind turbine structures is the characterisation of crack initiation and growth behaviour in large-scale monopile welded structures. The monopiles, which are made of thick-walled steel plates welded to each other in longitudinal and circumferential directions, are subjected to severe cyclic loading conditions in the harsh offshore environment. Therefore, the fatigue crack growth (FCG) behaviour of these welded structures needs to be characterised both in air and in seawater. This paper describes the procedure to extract standard size compact tension, C(T), specimens from the base metal (BM), heat affected zone (HAZ) and weld metal (WM) sections of double V-groove S355 welded plates and testing the specimens under cyclic loading conditions. The obtained data from these experiments have been analysed and the fatigue crack growth rates are correlated with the linear elastic fracture mechanics parameter, ΔK, to characterise the crack growth behaviour of each material (i.e. BM, HAZ and WM) in near threshold and Paris regions. The FCG trends obtained from these tests are discussed in terms of the material microstructure, mechanical properties and welding residual stress effects on life prediction of offshore wind turbine monopile structures.

[1]Offshore Renewable Energy Engineering Centre, School of Water, Energy and Environment, Cranfield University, Cranfield, MK43 0AL, UK
[2]Centrica Distributed Energy & Power, Millstream East, Maidenhead Road, Windsor SL4 5GD, UK

7. FATIGUE AT WELDS AND NOTCHES I

DIFFERENT APPROACHES FOR FATIGUE ASSESSMENT OF BUTT-WELDED JOINTS

C. Mario Rizzo[1], F. Berto[2] & T. Welo[2]

In recent years, fatigue assessment approaches in between fracture mechanics concepts and more traditional stress based methods have been proposed. Considering parameters describing the singular stress field at the notch tip where fatigue cracks initiate, the Notch Stress Intensity Factor (N-SIF) concept, introduced by Gross and Mendelson in 1972 following the work of 1952 by Williams, has been recently suggested as the fatigue governing parameter by Lazzarin and co-workers. Being N-SIF rather cumbersome for numerical estimations, other related quantities have been proposed and the corresponding fatigue assessment approaches developed. Among the others, the Strain Energy Density (SED) concept appear well established, being available a number of applications presented in open literature.

The above mentioned approach seems to overcome the difficulties introduced by other local fatigue assessment approaches, accounting for different effects not considered by methods based on a conventional stress range value. For example, observing that a based energy parameter depends on all loading modes, particularly the SED criterion can be successfully used to estimate the fatigue strength of welded details subjected to multiaxial fatigue loading. Also, automation of finite element method analyses seems possible by adopting dedicated sub-modelling techniques and re-meshing routines. Such significant features may allow the screening of complex structures with the aim to identify and assess fatigue critical details.

Butt joints are spread in all ship and offshore structures as well as in a number of welded assemblies, involving plate thicknesses ranging from few millimetres up to one hundred and even more. These joints are an appropriate test case to evaluate the potentially wide-ranging applications of N-SIF based fatigue assessment approaches mentioned above.

The main aim of this work is the fatigue strength assessment of a butt joint with weld reinforcement, typically used in steel works and especially in shipbuilding applications. A sensitivity analysis has been carried out by varying different geometric parameters of the welded joints.

Moreover, two different approaches for fatigue strength estimates were applied, namely the Effective Notch Stress Approach, which is nowadays widely used and it is supported by comprehensive literature and the Strain Energy Density approach (SED), the latter one newly introduced in scientific literature and still being tested within industrial practice.

As a comparison term, analyses were repeated directly evaluating the N-SIF value at the notch tip to obtain target values to assess the results obtained from the other approaches. Analyses were carried out by using two different finite element codes: Ansys® and Adina®. The application of different FEM software allowed verifying the quality of the obtained results and disclosed certain essential aspects related to the extrapolation of values from integration points of finite elements.

N-SIF based approaches are not yet recognised for fatigue life assessment in the various industrial fields nor were they agreed to be included in any regulation or standard. Though, the present study highlights the prospective capabilities of such approaches as well as it outlines modelling strategies and best practices to achieve consistent results.

[1]University of Genova, DITEN – Marine Structures Testing Lab, Via Montallegro 1, 16145 Genova (Italy)
e-mail: cesare.rizzo@unige.it
[2]NTNU, Department of Engineering Design and Materials, Richard Birkelands vei 2b, 7491 Trondheim (Norway)
e-mail: filippo.berto@ntnu.no; torgeir.welo@ntnu.no

FATIGUE ASSESSMENT OF WELDED JOINTS IN LARGE STEEL STRUCTURES: A MODIFIED NOMINAL STRESS DEFINITION

M. Colussi*[1], F. Berto[2] & G. Meneghetti[1]

According to the notch stress intensity factor (N-SIF) approach to the fatigue assessment of welded joints, the weld toe and the weld root are modeled as sharp V-notches (zero radius). The N-SIFs quantify the intensity of the elastic stress fields near the points of singularity (weld toe and root) and capture not only the weld shape effect, but also the size and the loading condition (membrane and bending) effects, so that the fatigue strength of many steel welded joints follows the same design curve. The computational effort required to evaluate the N-SIFs, due to the necessity of very refined meshes, is a strong limitation in industrial applications, especially in three-dimensional modeling of large structures. The strain energy density (SED), averaged over a control volume which embraces the weld root and the weld toe, keeps the robustness of the N-SIFs, allowing meshes with a grade of refinement up to three orders of magnitude larger. However, it remains a difficult challenge to perform massive fatigue investigations in large structures, whose only possibility of numerical representation is often based on really coarse and shell based finite element models. In this context, today a concrete way to address the problem, where concrete means complying with standards in force and nowadays design offices computational capabilities, can be found on nominal stress based methods. Here, a modified nominal stress is defined and how to compute it by means of coarse finite element models is shown. The effect of loading condition (membrane and bending) is also addressed by introducing a proper coefficient obtained by coupling local approaches (N-SIF and SED) advantages to the classical nominal stress components. Then, the implications in fatigue design of large steel structures are discussed.

*Corresponding author: marco.colussi.1@phd.unipd.it
[1]University of Padua, Department of Management and Engineering, Stradella S. Nicola 3, 36100 Vicenza, Italy

[2]NTNU, Department of Engineering Design and Materials, Richard Birkelands vei 2b, 7491 Trondheim, Norway

AVERAGED STRAIN ENERGY DENSITY-BASED SYNTHESIS OF CRACK INITIATION LIFE OF NOTCHED TITANIUM AND STEEL BARS UNDER UNIAXIAL AND MULTIAXIAL FATIGUE

A. Campagnolo[1], G. Meneghetti[1], F. Berto[2] & K. Tanaka[3]

The fatigue behaviour of circumferentially notched specimens made of titanium grade 5 alloy, Ti-6Al-4V, and austenitic stainless steel, AISI 304L, has been analysed.

With the aim of investigating the effect of the notch geometry and the loading condition on the fatigue strength of the titanium alloy, pure bending, pure torsion and multiaxial bending-torsion fatigue tests have been carried out on specimens characterised by two different root radii, namely 0.1 and 4 mm. As regards the notched steel bars, only the notch effect on the torsion fatigue behaviour has been analysed in detail by testing specimens characterised by three different notch tip radii: 0.1, 1 and 4 mm. In all cases, the nominal load ratio R has been kept constant and equal to -1.

Crack nucleation and subsequent propagation have been accurately monitored by using the electrical potential drop technique, which allows to define the crack initiation life in correspondence of an increase of the electrical potential drop.

The experimental fatigue results of each material have been re-analysed using the local strain energy density (SED) averaged over a control volume having radius R_0 surrounding the notch tip. With the aim of excluding all extrinsic mechanisms acting during the fatigue crack propagation phase, such as sliding contact and/or friction between fracture surfaces recently discussed by Tanaka dealing with the torsional fatigue behaviour of notched steel bars, the crack initiation life has been considered in the present work. The control radius R_0 for fatigue strength assessment of notched components, thought of as a material property, has been estimated for the two considered materials by imposing the constancy of the averaged SED for both smooth and cracked specimens at $N_A = 2$ million cycles. Therefore, R_0 is seen to depend on two material properties: the plain material fatigue limit (or the high-cycle fatigue strength of smooth specimens) and the threshold value of the stress intensity factor (SIF) range for long cracks.

[1]University of Padova, Department of Industrial Engineering, Via Venezia 1, 35131, Padova (Italy)
[2]NTNU, Department of Engineering Design and Materials, Trondheim (Norway)
[3]Department of Mechanical Engineering, Meijo University, 468-8502, Nagoya (Japan)

8. MULTIAXIAL FATIGUE I

RECENT DEVELOPMENTS ON MULTIAXIAL FATIGUE STRENGTH OF TITANIUM ALLOYS

F. Berto[1]*, A. Campagnolo[2] & S. Vantadori[3]

An accurate review of some recent data on multiaxial fatigue strength of severely notched titanium grade 5 alloy (Ti-6Al-4V) is carried out in the present paper. Experimental tests under combined tension and torsion loading, both in-phase and out-of-phase, have been carried out on axisymmetric V-notched specimens considering different nominal load ratios (R = -1, 0, 0.5). All specimens are characterized by a notch tip radius less than 0.1 mm, a notch depth of 6 mm and a notch opening angle equal to 90 degrees. The experimental data from multiaxial tests are compared with those from pure tension and pure torsion tests on un-notched and notched specimens, carried out at load ratio ranging from R = -3 to R = 0.5. In total, more than 160 fatigue data are re-examined, first in terms of nominal stress amplitudes referred to the net area and then in terms of the local strain energy density averaged over a control volume surrounding the V-notch tip. The dependence of the control radius on the loading mode is analysed showing a very different notch sensitivity for tension and torsion. For the titanium alloy Ti-6Al-4V, the control volume is found to be strongly dependent on the loading mode.

[1]NTNU, Department of Engineering Design and Materials, Richard Birkelands vei 2b, 7491, Trondheim, Norway
[2]University of Padova, Department of Industrial Engineering, Padova (Italy)
[3]Dept. of Civil-Environmental Engineering and Architecture, University of Parma, Parma – Italy
* Email: berto@gest.unipd.it

ACCURACY OF SOFTWARE MULTI-FEAST© IN ESTIMATING MULTIAXIAL FATIGUE DAMAGE IN METALLIC MATERIALS

L. Susmel[1]

This paper investigates the accuracy and reliability of software Multi-FEAST© (www.multi-feast.com) in estimating fatigue strength of metallic materials subjected to time-variable multiaxial load histories. Multi-FEAST©'s core algorithm is based on the Modified Wöhler Curve Method which is a bi-parametrical critical plane approach assuming that Stage I fatigue cracks initiate on those material planes experiencing the maximum shear stress amplitude. In the present investigation, the overall accuracy of Multi-FEAST© was checked against a large number of experimental results taken from the literature and generated by testing metallic specimens under constant and variable amplitude multiaxial fatigue loading. Such a comprehensive validation exercise allowed us to demonstrate that Multi-FEAST© is highly accurate and reliable. This confirms that this software is a powerful design tool suitable for estimating multiaxial fatigue damage in situations of practical interest.

[1]Department of Civil and Structural Engineering, The University of Sheffield, Sheffield, S1 3JD, UK – e-mail: l.susmel@sheffield.ac.uk

MULTIAXIAL LOW CYCLE FATIGUE ANALYSIS OF NOTCHED SPECIMENS FOR TYPE 316L STAINLESS STEEL UNDER NON-PROPORTIONAL LOADING

T. Morishita[1], S. Bressan[2], T. Itoh[1], P. Gallo[3] & F. Berto[4]

This study re-analyzes some multiaxial low cycle fatigue life of notched specimen under proportional and non-proportional loadings at room temperature. Strain controlled multiaxial low cycle fatigue tests were carried out using smooth and circumferentially notched round-bar specimens of type 316L stainless steel. Four kinds of notched specimens were employed with elastic stress concentration factors, K_t, as 1.5, 2.5, 4.2 and 6.0. The strain paths include proportional and non-proportional loadings. The former employed was a push-pull strain and a reversed torsion strain. The latter was achieved by strain path where axial and shear strains has 90° phase difference but their amplitudes are the same based on von Mises' criterion. The notch dependency of multiaxial low cycle fatigue life and the life estimation are discussed with employing inelastic finite element analysis. The fatigue life depends on both K_t and strain path. The strain parameter for the life estimation is discussed with the non-proportional strain parameter proposed by Itoh et al. with introducing K_t. The proposed parameter gave a satisfactory evaluation of multiaxial low cycle fatigue life for notched specimen of type 316L stainless steel under proportional and non-proportional loadings. In addition, a new predictive model has been proposed taking into account the real strain hardening behavior of the material in the proximity of the notch tip. The new model allows to evaluate the data in a narrow scatter band and provides a sound interpretation of the crack initiation phase. The new model has been compared with the original parameter proposed by Itoh et al.

[1]Ritsumeikan University, Department of Mechanical Engineering, College of Science & Engineering, 1-1-1, Nojihigashi, Kusatsu-shi, Shiga, 5258577, Japan
[2]Ritsumeikan University, Graduate School of Science & Engineering, 1-1-1,

Nojihigashi, Kusatsu-shi, Shiga, 5258577, Japan
³Aalto University, Department of Mechanical Engineering, Marine Technology, Puumiehenkuja 5A, 02150 Espoo, Finland
⁴NTNU, Department of Engineering Design and Materials, Richard Birkelands vei 2b, 7491, Trondheim, Norway

EFFECT OF INTERCHANGE IN PRINCIPAL STRESS AND STRAIN DIRECTIONS ON MULTIAXIAL FATIGUE STRENGTH OF TYPE 316 STAINLESS STEEL

Y. Takada[1], T. Morishita[1] & T. Itoh[2]*

Machines and structures in service conditions undergo multiaxial fatigue damage under proportional and non-proportional loading conditions rather than uniaxial damage. The non-proportional loading is defined as the loading in which principal stress and strain directions change during a cycle. Under the non-proportional loading, it has been reported that the fatigue lives are reduced accompanying with additional hardening depending on material and loading history. Some researchers suggest that the fatigue lives are reduced by the loading in which the directions of maximum amplitude of principal stress and strain are interchanged into two directions due to the direction change of maximum shear stress and strain planes, even if the loading is the proportional loading. However, it is still in an open issue that the effect of the direction change on the fatigue strength due to lack of experimental results.

In this study, multiaxial fatigue tests were carried out using a hollow cylinder specimen of type 316 stainless steel to investigate the effect of interchange in the directions of maximum shear stress and strain planes. An electric servo control multiaxial fatigue testing machine, which can apply combined push-pull, reversed torsion and cyclic inner pressure, was used. Loading patterns were the proportional loading and the non-proportional loading. The former is the push-pull and the latter is the combined push-pull, reversed torsion and inner pressure in which directions of maximum shear stress and strain planes are interchanged with the progress of the cycle. The effect of the direction interchange on the fatigue strength was discussed based on the experimental results.

[1]Graduate School of Science & Engineering, Risumeikan University
[2]Department of Mechanical Engineering, College of Science & Engineering,

Ritsumeikan University, 1-1-1, Noji-higashi, Kusatsu-shi, Shiga 525-8577, Japan
*Corresponding author: Tel.: +81 (0)77 561 4965; Fax: +81 (0)77 561 2665.
E-mail: itohtaka@fc.ritsumei.ac.jp (T. Itoh)

9. FATIGUE OF ENGINEERING MATERIALS

FATIGUE PROPERTIES OF SOLUTION STRENGTHENED FERRITIC DUCTILE CAST IRONS IN HEAVY SECTION CASTINGS

T. Borsato[1], F. Berto[2], P. Ferro[1] & C. Carollo[3]

Recently, standardized solution strengthened ferritic ductile cast irons (SSF-DI) have met the interest of the industrial world due to their improved mechanical properties and workability compared to standard ferritic-pearlitic ductile cast irons. However, the limited number of experimental data and the lack of production experience make the introduction of these new alloys to the market very difficult. For this reason, mechanical and fatigue properties of heavy section SSF-DI castings have been examined. Metallographic analyses have been performed using optical microscopy to identify the most important microstructural parameters. Fracture surfaces of fatigue specimens have been investigated using a Scanning Emission Microscope in order to identify crack initiation and propagation zones. Fatigue curves of SSF-DI have been finally compared with those obtained from traditional ductile cast iron specimens taken from the same casting geometry.

[1]University of Padova, Department of Engineering and Management, Stradella S. Nicola, 3 I-36100 Vicenza, Italy.
[2]NTNU, Department of Engineering Design and Materials,Richard Birkelands vei 2b, 7491, Trondheim – Norway.
[3]VDP Fonderia SpA, via lago di Alleghe 39, 36015 Schio, Italy.

CONTACT FATIGUE CARACHTERIZATION OF 17NiCrMo6-4 SPECIMENS USING A TWIN-DISC TEST BENCH

A. Terrin[1] & G. Meneghetti[2]*

In structural design of power transmission systems, the evaluation of the load carrying capacity of gears with regard to contact fatigue still remains a tricky issue. According to ISO 6336 recommendations, design against pitting is based on Hertzian contact stresses, whose limit values should preferably be derived experimentally using meshing gears. However, the ISO standard suggests also to derive material characteristic values by means of simpler and faster tests on rolling pair of disks in loaded contact, such tests being particularly suitable to compare the contact fatigue behavior of either different materials or manufacturing processes. This approach results in a considerable cost saving and allows a rapid comparison between different materials and treatments with specimens that are easy to manufacture. In this work, a new twin-disc test rig was developed for the contact fatigue characterization of gear materials. The specimen were designed to recreate the meshing conditions between sun and planet gears of the planetary gear set located in the wheel hub of off-highway vehicles. Sun gears indeed, are particularly prone to manifest pitting damages because of the high frequency of contacts due to the meshing with three or four planets, and because of the small teeth dimensions which results in high contact pressures even with moderate applied loads. Pitting usually origins at the dedendum of the sun gear, therefore the curvature radii and sliding velocity at this location, calculated for a medium power axle for agricultural applications, were recreated in the tests through an adequate geometry of the specimens. In order to minimize the need for in-situ monitoring by an operator and to rely on a uniform end-of-test criterion, the twin-disc test rig was equipped with a vision system conceived to automatically identify the presence of pits on the surface of the specimens.
A series of tests were performed on 17NiCrMo6-4 discs with two different slide-to-roll ratios. Disc specimens with the same slide-to-roll ratio of the sun gears showed a similar damaging mode, but a 33% higher load capacity.

[1]PhD student at the University of Padova, Department of Industrial Engineering, Via Venezia 1- 35131- Padova (Italy), and R&D department Carraro S.p.A., Via Olmo 37 - 35010 - Campodarsego, PD (Italy)
e-mail: andrea.terrin@studenti.unipd.it
[2]Associate Professor in Machine Design, - University of Padova, Department of Industrial Engineering, Via Venezia 1 – 35131 - Padova (Italy).
e-mail: giovanni.meneghetti@unipd.it

OVERVIEW OF COPPER WIRE FAILURE MECHANISMS IN ROTATING MACHINERY

H. Y. Ahmad[1], D. Bonnieman[2] & D. C. Howard[3]

This paper is an outline study of fatigue failure of a tough pitch copper wire brazed to a silver plated beryllium copper strip using Sil-Fos filler materials. This paper discusses the effect of the brazing manufacturing process conditions; such as temperature, pressure and filler material flow in the joint around the copper wire; on the fatigue lifetime of the copper wire. Examination of the fracture surface of copper wire showed a hydrogen embrittlement effect which is triggered by the brazing process environment and potential stress concentrations resulting from the manufacturing process. Finite element analysis was conducted to determine the effect of the formed wire radii of the copper wire when brazed to beryllium copper strips and subjected to thermal cyclic loads and centripetal force.

[1]Engineering Consultant, Stress & Materials, Safran Electrical & Power, Pitstone, Buckinghamshire, UK. e-mail: hayder.ahmad@Safrangroup.com
[2]Engineering Consultant, Stress & Materials, Safran Electrical & Power, Pitstone, Buckinghamshire, UK. e-mail: david.bonnieman@safrangroup.com
[3]Engineering Consultant, Machine Design, Safran Electrical & Power, Pitstone, Buckinghamshire, UK. e-mail: Darren.howard@safrangroup.com

LOW CYCLE FATIGUE BEHAVIOUR OF AN ULTRAFINE-GRAINED METASTABLE AUSTENITIC CrMnNi STEEL

M. Droste*[1], M. Fleischer[1], A. Weidner[1] & H. Biermann[1]

An ultrafine grained (UFG) metastable TRIP steel (16.6Cr-7.1Mn-6.4Ni-0.05C-0.02N) obtained by swaging and subsequent reversion annealing has been investigated and compared to its coarse-grained (CG) counterpart (16.1Cr-6.0Mn-6.0Ni-0.04C-0.04N). The materials exhibited mean grain diameters of 0.7 μm and 14 μm, respectively. This study focusses on their low cycle fatigue behaviour including the fatigue induced martensitic phase transformation, which occurred for both grain sizes. For this purpose total strain controlled fatigue tests with amplitudes in the range of $0.3\,\% \leq \Delta\varepsilon_t/2 \leq 1.2\%$ have been conducted at room temperature.

The yield stress, the cyclic stress as well as the fatigue life are clearly increased for the UFG material. Concerning the α'-martensite formation, which was recorded during the tests using a ferrite sensor, the amounts at the end of fatigue life are similar for the high strain amplitudes, while the medium and small strain amplitudes lead to higher amounts in case of the CG material. However, the impact on the stress response curves in terms of a strong cyclic hardening due to the α'-martensite formation was much less pronounced for the UFG steel, which nevertheless exhibited higher cyclic stresses due to the Hall-Petch effect.

Furthermore, BSE (backscattered electrons) and EBSD (electron backscatter diffraction) measurements have been performed on specimens after reversion annealing as well as in the cyclically deformed state for selected strain amplitudes in order to investigate the microstructure evolution regarding grain sizes, phase fractions and deformation mechanisms.

[1]Technische Universität Bergakademie Freiberg, Institute of Materials, Engineering, Gustav-Zeuner-Straße 5, 09599 Freiberg, Germany
*Corresponding author: Matthias.Droste@iwt.tu-freiberg.de

10. FATIGUE AT WELDS AND NOTCHES II

SIGNIFICANCE OF UNDERCUTS ON FATIGUE STRENGTH OF BUTT-WELDED JOINTS

C. Steimbreger & M. D. Chapetti

Undercut formation affects fatigue strength of welded joints, since it constitutes a notch at the weld toe. The importance of these defects in welded structures demands accurate prediction of fatigue lifetime. Detrimental effects of undercuts are governed by stress concentration, which is characterized by notch depth and root radius. From this standpoint, numerical simulations of transversely stressed butt-joints were performed. Relationship between undercut geometry and fatigue strength of welds were studied by means of a fracture mechanics approach. Relative influence of other involved parameters such as size of defects, plate thickness and type of loading, were also analyzed.

Laboratory of Experimental Mechanics (LABMEX), INTEMA (Institute for Material Science and Technology), CONICET - University of Mar del Plata
J.B. Justo 4302, (B7608FDQ) Mar del Plata, Argentina
mchapetti@fi.mdp.edu.ar

CONSIDERATION OF WELD DISTORTION THROUGHOUT THE IDENTIFICATION OF FATIGUE CURVE PARAMETERS USING MEAN STRESS CORRECTION

Y. Gorash[1,*], X. Zhou[1,], T. Comlekci[1,], D. MacKenzie[1] & J. Bayyouk[2]

Weld distortion is an essential attribute of a mechanical testing specimen, where two parts are joined with a butt weld. It needs to be addressed when doing tensile and, particularly, fatigue testing because clamping of the specimen in the testing machine usually induces an additional component of bending stress. This can be considered as a constant or mean stress in the experiment, while the membrane stress component is variable. Moreover, in the case of significant distortion over 2°, the bending stress can be greater than the membrane stress range. Therefore, the influence of bending stress on fatigue life should be considered by the introduction of a mean stress correction into the fatigue curve fitting procedure.

The experiments comprised fatigue testing of weld specimens with the target stress ratio R = 0 until the failure defined by separation. The weld specimens were made of a moderately strong weldable structural steel equivalent to BS 4360 grade 50D with a yield point of about 415 MPa and tensile strength of about 595 MPa. The specimens had a typical shape according to ISO/TR 14345:2012, but a stepped thickness, since they were cut using a water jet cutter from two butt welded plates having two different thicknesses (0.5 in and 0.625 in). The testing was performed at the frequency of 10Hz using 15 samples (5 load levels, 3 samples each) for the stress ratio R = 0.1 and the stress amplitude varying from 60 MPa to 110 MPa. The specimens had a random angular distortion varying from 0.3° to 2.5° measured optically, which was verified by strain gauge measurements when gripped by the testing machine and confirmed by the structural FEA results of corresponding specimens geometry incorporating distortion.

A generated S-N curve based on the obtained experimental data is intended for use with the fatigue post-processor nCode DesignLife for the fatigue life predictions. This software requires the input of S-N curves described by a power-law equation, which is an inverted form of Basquin's equation

[1]. The mean stress correction used in fatigue analysis of welds is applied through the FKM approach [2]. The experimental data is presented in 3D-space of stress range, cycles and mean stress ($\Delta\sigma$, N, σm). This data is fitted by a surface defined using the function for $\Delta\sigma$, which combines Basquin equation and FKM correction, resulting in a non-linear dependence on N and linear on σm.

This paper focuses on the discussion of the 2-step procedure for the identification of the required parameters of the fitted surface. As a result, a conversion expression is obtained for the stress range intercept parameter, which can be used to derive S-N curves for any value of R. After application of the mean stress and thickness normalisation, the obtained weld S-N curve is compared to the generic weld S-N curves provided in the material database of nCode DesignLife, and the difference between them is pointed out.

References:
[1] Basquin O.H. The Exponential Law of Endurance Tests, In: Proc. ASTM, Vol.10, ASTM: Philadelphia, PA, 1910, pp. 625-630
[2] FKM (Forschungskuratorium Maschinenbau), FKM-Guideline: Analytical Strength Assessment of Components in Mechanical Engineering, VDMA Verlag: Frankfurt / Main, 5th edition, 2003

[1]Dep. of Mechanical & Aerospace Engineering, University of Strathclyde, Glasgow G1 1XJ, UK
[2]Weir Oil & Gas, Weir SPM, Fort Worth, TX 76108, USA
*Corresponding author: yevgen.gorash@strath.ac.uk

CALCULATION OF STRESS CONCENTRATION FACTORS IN OFFSHORE WELDED STRUCTURES USING 3D LASER SCANNING TECHNOLOGY AND FINITE ELEMENT ANALYSIS

L. Wang[1]*, W. Vesga Rivera[1], A. Mehmanparast[1], F. Brennan[1], A. Kolios[1], I. Tavares[2]

Offshore wind turbine support structures (such as monopiles) generally consist of several steel cans, with adjacent cans generally connected together through welded joints. One of the most important factors to evaluate the localized increase in stress which depends on the quality of the welded joints is the stress concentration factor (SCF). The complex welding profiles in offshore structures makes the accurate calculation of SCF quite challenging. In this work, a novel approach for the calculation of SCFs in offshore welded structures is proposed which consists of combining three-dimensional (3D) laser scanning technology (LST) and finite element analysis (FEA). The high-precision geometry of the as-built welded specimens is obtained using 3D LST, and then imported into a finite element software package to perform FEA modelling to calculate SCFs. The proposed approach is applied to calculate SCFs in large-scale S-N fatigue welded specimens tested during the Structural Lifecycle Industry Collaboration (SLIC) joint industry project. The variation of SCFs along the width of the welded specimens is shown and compared to identify the range of SCFs in actual test specimens. The critical point where fatigue crack is most likely to initiate is also identified.

[1]Centre for Offshore Renewable Energy Engineering, School of Water, Energy and Environment, Cranfield University, Cranfield, MK43 0AL, UK.
[2]SLIC Project Manager, Centrica Renewable Energy Limited, Windsor, UK.

HIGH-RESOLUTION 3D ASSESSMENT AND ACPD CRACK GROWTH MONITORING OF WELD TOE FATIGUE CRACK INITIATION

S. Chaudhuri*[1], J. Crump[2], P. A. S. Reed[1], B. G Mellor[1], P. Tubby[2], M. Waitt[3] & A. Wojcik[3]

Weld toe fatigue crack initiation is highly dependent on local geometrical flaws such as inclusions, undercuts and cold-laps. These features are known to promote initiation early in the fatigue life and the weld profile is thought to influence how these small micro-cracks evolve. Understanding the influence of the local weld toe geometry on early fatigue crack growth is still fairly limited. It is however important particularly in the high-cycle regime where minimising critical geometries has the potential to maximise fatigue lives.

In this work short cracks (<500µm) have been initiated and interrupted on non-load carrying welds manufactured with different electrodes. A multiple probe high-resolution alternating current potential drop method (ACPD) method was used to capture the earliest stages of crack initiation via a change in resistance. Fatigue crack initiation occurred earliest in the welds manufactured using metal-core electrodes. The rate of PD change also suggesting early growth was faster in these welds. These results compliment S-N data of the same welds in which welds manufactured using the solid-core electrodes had longer lives in the high-cycle regime.

To understand further the influence of all parameters on the early fatigue life: hardness measurements, residual stress measurements and micro-computed tomography (µ-CT) has been used to fully characterise the welds. A novel work flow has been developed which creates 3D finite element models from the 3D µ-CT data. This enables a more accurate assessment of local weld toe stresses which for the first time can be correlated directly with the earliest stages of fatigue crack initiation.

[1]Engineering Materials Research Group, Faculty of Engineering and the Environment, University of Southampton, UK
[2]TWI Ltd, Cambridge, UK

[3]Matelect Ltd, UK
*Corresponding author email address: sc24g14@soton.ac.uk

11. MULTIAXIAL FATIGUE II

MULTIAXIAL FATIGUE TEST AND STRENGTH USING CYCLIC BENDING AND TORSION TESTING MACHINE

H. Sato[1], T. Morishita[1], T. Itoh[2]* & M. Sakane[2]

Actual components and structures undergo multiaxial fatigue damage rather than uniaxial damage. Cyclic multiaxial loading sometimes becomes non-proportional loading in which principal stress and strain directions change during a cycle. Several researchers have reported that effects under non-proportional loading in high and low cycle fatigue regions. However, a reduction of fatigue limit under non-proportional loading is still open question which is reasoned from that high cycle fatigue tests require much longer time compared to low cycle fatigue tests under non-proportional loading condition. Itoh and Morishita et al. developed a multiaxial high cycle fatigue testing machine to perform the fatigue test with a high frequency up to 50Hz under the non-proportional loading. Two types of loading patterns were employed: a cyclic bending loading (proportional loading) and a combined cyclic bending loading and reversed torsion loading with a 90 degree phase shift (non-proportional loading).
In this study, high cycle fatigue tests under proportional and non-proportional loading conditions were performed to evaluate fatigue strength. The utility of the developed testing machine was also confirmed.

[1]Graduate School of Science & Engineering, Ritsumeikan University,
[2]Department of Mechanical Engineering, College of Science & Engineering, Ritsumeikan University
1-1-1, Noji-higashi, Kusatsu-shi, Shiga 525-8577, Japan
*Corresponding author: Tel.: +81 (0)77 561 4965; Fax: +81 (0)77 561 2665.
E-mail: itohtaka@fc.ritsumei.ac.jp (T. Itoh)

NUMERICAL STUDY ON INFLUENCE OF NON-PROPORTIONAL STRESSING ON FRETTING FATIGUE LIFETIME ASSESSMENT

R. H. Talemi

Fretting fatigue failure results from oscillatory micro-slip between two contacting parts which are subjected to cyclic fatigue loading at the same time. This phenomenon may occur in many applications such as bolted and riveted connections, steel cables, bearings shafts, steam and gas turbines. The fretting fatigue may reduce the lifetime of a component by half or even more, in comparison to the plain fatigue.

Fretting fatigue life assessment under multiaxial loadings is a rather complex task, as the stresses at contact interface are multiaxial. Structural components such as a steel bolted connection subjected to multiaxial fretting fatigue loadings can respond differently depending on the non-proportionality of the stress at contact interface. A phase-shift between the acting stress components results in a reduction of lifetime for ductile materials. Furthermore, the introduction of a phase-shift between stress components leads to lower equivalent stress values calculated by most of available multiaxial fatigue criteria and therefore to an overestimated fatigue life.

In this study an external non-proportionality factor is introduced to predict fretting fatigue lifetime under non-proportional stressing. The non-proportionality of the stress implies that time-dependent normal and shear stress components in different planes are essentially of different shapes. The proposed non-proportionality factor is based on a statistical correlation of stress components. To this end, a two dimensional finite element model was used to extract the stress components at the contact interface. Moreover, to enhance fretting fatigue lifetime assessment, a stress-based multiaxial fatigue criterion was modified by introducing the non-proportionality factor. Finally, the results were compared against a series of experimental data taken from literature.

ArcelorMittal Global R&D Gent-OCAS N.V., Pres. J.F. Kennedylaan 3, 9060 Zelzate, Belgium
Email: reza.hojjatitalemi@arcelormittal.com

THE HEAT ENERGY DISSIPATED IN A CONTROL VOLUME TO CORRELATE THE FATIGUE STRENGTH OF SEVERELY NOTCHED AND CRACKED STAINLESS STEEL SPECIMENS

G. Meneghetti* & M. Ricotta

In the last years the authors proposed and developed a heat energy-based approach able to rationalise the fatigue behaviour of plain and bluntly notched specimens made of AISI 304L stainless steel subjected to constant amplitude and two load level fatigue tests. The mean stress influence in fatigue has also been taken into account. The proposed approach assumes the heat energy dissipated in a unit volume of material per cycle, Q, as a fatigue damage index, which can be readily evaluated by means of temperature measurements performed at the crack initiation point. Nevertheless, Q can hardly correlate fatigue test results generated from severely notched specimens, because the well-known notch support effect makes questionable the use of peak quantities (stress-, strain- or energy-based) evaluated at the notch tip. Rather, the fatigue-relevant quantity (either stress-, strain- or energy-based) should be averaged inside a material dependent structural volume.

Recently, the authors presented an experimental technique able to evaluate the heat energy dissipated in a small material volume surrounding the tip of a fatigue crack, Q*. Such a technique is based on a theoretical model and on the radial temperature profiles measured from the crack tip outward by means of an infrared camera. The aim of the present paper is to use the parameter Q* to correlate the fatigue strength of severely notched and cracked specimens. To this end, tension-compression axial fatigue tests were conducted on specimens machined from 4-mm-thick AISI 304L steel plate. Q* has been evaluated starting from temperature profiles measured by means of a FLIR SC7600 infrared camera.

University of Padova, Department of Industrial Engineering, Via Venezia 1, 35131 Padova, Italy
*Corresponding author: Prof. G. Meneghetti

12. INFLUENCE OF MANUFACTURING PROCESSES ON FATIGUE

HIGH CYCLE FATIGUE BEHAVIOUR IN ULTRASONIC-SHOT-PEENED B-TYPE Ti ALLOY AND EBSD-ASSISTED FRACTOGRAPHY OF SUB-SURFACE FATIGUE CRACK INITIATION

Y. Uematsu[*1], T. Kakiuchi[1] & K. Hattori[2]

Rotating bending fatigue tests had been conducted using ultrasonic-shot-peened (USPed) β-type Ti alloy Ti-22V-4Al. As-received and solution-treated (ST) materials were also used in the fatigue tests for comparison. As-received material had α+β dual phase, where EBSD analysis revealed that the area fraction of β phase was around a few percent. In contrast, ST specimen has full β microstructure. As-received specimens exhibited higher fatigue strengths than ST ones due to the presence of hard α phase. Subsequently, ultrasonic shot peening (USP) was applied to the as-received material. Fatigue strengths were highly improved by USP in the finite life region (10^5~10^7 cycles). However, the fatigue strengths of USPed and as-received specimens were nearly comparable in very high cycle regime around 10^8 cycles. The similar fatigue strengths in very high cycle regime were attributed to the sub-surface crack initiation in the USPed specimens. In general, sub-surface cracks initiate from intermetallic compound inclusions in high strength steels. But the fractographic analyses revealed that inclusion was not recognized at the sub-surface crack initiation site in the USPed β-type Ti alloy. Thus EBSD analyses were performed near the sub-surface crack initiation site. It was found that α-phase rich grain was the sub-surface crack initiation site, indicating that α-phase rich grain could be microstructurally weak point in β-type Ti alloy. In this research, the beneficial effects of USP on the fatigue performances in β-type Ti alloy were clearly shown. Furthermore, sub-surface crack initiation mechanism without inclusion in β-type Ti alloy was proposed based on EBSD phase analyses.

*Corresponding author: Tel: +81-58-293-2501, E-mail: yuematsu@gifu-u.ac.jp
[1]Gifu University, Department of Mechanical Engineering, 1-1 Yanagido, Gifu 501-1193, Japan

[2]Toyoseiko Co.,Ltd., 3-195-1 Umaganji Yatomi-City, Aichi, 490-1412, Japan

EFFECT OF ADDITIVE ELEMENTS ON CRACK INITIATION AND PROPAGATION BEHAVIOUR FOR LOW-Ag SOLDERS

N. Hiyoshi*[1], M. Yamashita[2] & H. Hokazono[3]

This study discusses the effect of additive elements on crack propagation behavior for Sn1.0Ag0.7Cu low-Ag content lead-free solders at high temperature. Low-Ag content solder is expected to reduce the forming of coarse intermetallic compound and to reduce material cost. Cyclic tension-compression fatigue test for four kinds of Sn1.0Ag0.7Cu solders with center holed specimen (Figure 1) were conducted at 313K in order to make clear the effect of additive elements on crack initiation and propagation behavior. Sn1.0Ag0.7Cu (SnAgCu) is the basis content, Sn1.0Ag0.7Cu0.07Ni0.01Ge (SnAgCuNiGe) is the solder which is added Ni and Ge to SnAgCu, Sn1.0Ag0.7Cu2.0Bi (SnAgCuBi) is the solder which is added Bi to SnAgCu, Sn1.0Ag0.7Cu2.0Bi0.07Ni0.01Ge (SnAgCuBiNiGe) is the solder which is added Bi, Ni and Ge to SnAgCu.

Stress amplitude of Bi-content solders were bigger than that of SnAgCu and SnAgCuNiGe (Figure 2). Crack initiation cycle of Bi-content solders were also earlier than that of SnAgCu or SnAgCuNiGe. Crack propagation cycle, which is calculated as subtract the crack initiation cycle from total number of cycle, was also considered in this study. SnAgCu solders with additive elements Bi have shorter crack propagation cycles than SnAgCu and NiGe added solder. This result indicate that the additive elements Bi have effects on crack initiation and propagation cycles, that is, the additive element Bi accelerate crack propagation rate. We had also discussed the adaptation of J-integral range parameter. J-integral range parameter evaluates the crack propagation rate for SnAgCu solders independent of the additive elements (Figure 3).

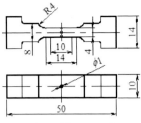

Fig. 1 Shape and
dimensions of the specimen
(mm).

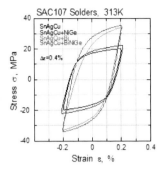

Fig. 2 Hysteresis loops of
solders.

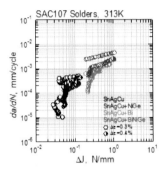

Fig. 3 J-integral evaluation
results.

[1]Department of Mechanical Engineering, University of Fukui
3-9-1 Bunkyo, Fukui-shi, Fukui 910-8507, Japan
[2]Fuji Electric Co., Ltd. 11-2, Ohsaki 1-chome, Shinagawa, Tokyo 141-0032,
Japan
[3]Fuji Electric Co., Ltd. 1, Fuji-machi, Hino-city, Tokyo 191-8502, Japan
*Corresponding author: hiyoshin@u-fukui.ac.jp, Tel&Fax: +81-776-27-9933

FATIGUE ASSESSMENT OF ADHESIVE WOOD JOINTS THROUGH PHYSICAL MEASURING TECHNIQUES

S. Myslicki[1], C. Winkler[2], N. Gelinski[1], U. Schwarz[2] & F. Walther[1]

Timber, as a structural material, has experienced its revival because of lightweight potential, excellent CO_2 emissions balance, high manufacturing efficiency, less transportation energy and potential to recycle. Adhesive bonding plays a major role in timber engineering because of its superiority over mechanical fasteners. Additionally, some structural timber product consists of a number of layers of timber bonded together with structural adhesives. Research on methods to design adhesively bonded timber joints under static loads is currently ongoing and significant progress has already been achieved.

Recently, it is observed that timber constructions for dynamic applications are establishing. Wind power plants with towers and blades out of timber, bridges and high-rise buildings are already realized. In future additional fields of applications, e.g. wood adhesively bonded with technical textiles, polymers or metal foils that may be used in automobile industry as a replacement or supplement to aluminum, magnesium or CFRP are possible. Therefore the fatigue characteristics of adhesive bonded wood must be investigated. Current literature to this topic is inconsistent and according to metal research the timber fatigue investigations are rare and unprogressive. The presented work studied the use of physical measurement techniques as the stress-strain hysteresis measurement to reveal fatigue characteristics, the acoustic emission to detect the spectrum of mechanical vibrations which gives information about the released energy due to fractures and the resistance of electrically conductive adhesive for the damage detection in the adhesive. These techniques were applied during the cyclic load increase approach which enables the estimation of the fatigue strength and to give insights of the damage mechanisms. With constant amplitude tests a good estimation of traditional S-N curve was possible. Additionally, the viscoelastic behavior of the wooden adhesive joints was observed. It was found, that the hysteresis opening, the dynamic modulus and the change in electrical resistance are good characteristic values for deformation and damage accumulation.

[1]TU Dortmund University, Department of Materials Test Engineering (WPT), Baroper Str. 303, D-44227 Dortmund, Germany
[2]Eberswalde University for Sustainable Development (HNEE), Faculty of Wood Science and Technology, Schicklerstraße 5, D-16225 Eberswalde, Germany
*Corresponding author: sebastian.myslicki@tu-dortmund.de

13. FATIGUE CRACK GROWTH AND CLOSURE II

EXPERIMENTAL INVESTIGATION OF THICKNESS EFFECTS ON FATIGUE CRACK CLOSURE BEHAVIOUR IN Al7075-T6 ALLOY

K. Masuda*[1], S. Ishihara[2] & M. Okane[2]

In this study, fatigue tests were conducted on aluminum alloy 7075-T6 to study fatigue crack growth and crack closure behavior of the material. The experiments included the determination of the crack-opening levels K_{op} as a function of ΔK, the rate of fatigue crack growth da/dN as a function of ΔK_{eff}, where $\Delta K_{eff}, (= K_{max} - K_{op})$ is the effective stress intensity factor range. In addition, a change in the value of K_{op} due to the specimen-surface removal and a crack front shape and fatigue striation were also investigated. The fatigue crack growth (FCG) tests were carried out in a servo controlled hydraulic test system under load control condition in laboratory air at room temperature under stress ratio of R = 0.1. The test frequency was 15 Hz. Crack lengths were measured using the replication technique. The elastic compliance method was used for a determination of the crack-opening level. It was concluded that Al7075-T6 exhibited the plasticity induced fatigue crack closure behaviour (PIFCC). The crack-opening level decreased with increasing the specimen thickness (1~21 mm). The extent of PIFCC behaviour of the material was decreased with an increase in the specimen thickness. The relation of da/dN and DKeff did not depend on specimen thickness. An effect of surface removal on the crack-opening level Kop was observed in each specimen thickness. After removal of the specimen surface, the crack-opening level dropped and then recovered to its original value while propagating. Due to the observation of crack front shape in each specimen thickness, it was found that the crack length at the surface was less than in the interior. It was concluded that the crack length at the specimen surface is less than in the interior due to higher crack-opening level Kop at the specimen surface than that at the specimen interior. A SEM (Scanning Electron Microscope) was also performed to investigate the striations for not only each specimen thickness but also the locations (near the specimen surface and the specimen interior).

[1]University of Toyama, 3190 Gofuku Toyama-shi Toyama, Japan
[2]National Institute of Technology, Toyama College, Japan
*Corresponding author: masuda@eng.u-toyama.ac.jp

THROUGH THICKNESS EVOLUTION OF CRACK TIP PLASTICITY

D. Camas[1], P. Lopez-Crespo[1], F. V. Antunes[2] & J. R. Yates[3]

Experimental methods to measure fracture mechanics parameters tend to provide information from or about the surface of cracked components. However, information about the interior of the component is key to understanding the mechanisms governing the damage processes at a crack tip for both fatigue and fracture events. In this work we present a detailed numerical analysis of the evolution of the plastic zone through the thickness of an aluminium alloy specimen. This is done by means of ultra-fine non-linear finite element models. The simulated results are compared with experimental displacement data measured optically from the surface of the specimen.

[1]Department of Civil and Materials Engineering, University of Malaga, C/Dr Ortiz Ramos, s/n, 29071 Malaga, Spain
[2]Department of Mechanical Engineering, University of Coimbra, Portugal
[3] Simuline Ltd., Derbyshire S18 1QD, UK

THE INVESTIGATION OF THE FATIGUE CRACK GROWTH RATE IN TRIPLE PHASE MICROSTRUCTURE OF A ROTOR STEEL

A. S. Golezani, M. Mobaraki & R. Samadi

In this study, the fatigue crack growth rates in the triple phase microstructure of ferrite-beinite-martensite (FBM) and microstructure of a tempered martensite (TM) were investigated. For this purpose the Crack-Tip Opening Displacement was used to measure the crack length at different cycles. The step quenching and quenching and tempering heat treatments produced TM and FBM microstructures, respectively. The crack length was measured according to the specification ASTM E647 and by methods of Saxena and Hudak. The fatigue crack growth rate in triple phase microstructure was slower than TM microstructure so that the amounts of m and A (are material constants in the Paris-Erdogan equation) for tempered martensite microstructure and FBM microstructure were calculated 3.51, 1.97x10-10 and 3.12, 1.41x10-11 (m/c), respectively. This can be attributed to the high number of the interphase boundaries in triple phase microstructure.

Department of Materials Science and Engineering, Karaj Branch, Islamic Azad University, Karaj, Iran.

14. FATIGUE OF COMPOSITE MATERIALS AND STRUCTURES

FATIGUE BEHAVIOUR AND MEAN STRESS EFFECT OF THERMOPLASTIC POLYMERS AND COMPOSITES

Z. Lu*, B. Feng & C. Loh

More and more polymers and polymer composite materials are used in automotive industry to reduce cost and weight of vehicles to meet the environmental requirement. However, the fatigue behaviour for these materials is less understanding than metallic materials. The current work is focussed on the fatigue behaviour for a range of thermoplastic polymer/composite materials. It reveals that the fatigue behaviour of these materials can be described by S-N curves using the Basquin Equation. All the materials exhibit significant mean stress effect. The most commonly used mean stress correction equations developed in metal fatigue were evaluated with the current test results. It reveals that Goodman, Gerber and Soderberg cannot be used as generic equations for the materials investigated, whereas Smith-Watson-Topper can correlate the test data reasonably well, but the best correlation is given by Walker with material constant $\gamma = 0.4$.

Jaguar Land Rover Ltd – Abbey Road, Whitley, Coventry CV3 4LF, UK
*Corresponding author. Tel: +44 24 76 204110
E-mail: zlu5@jaguarlandrover.com (Z. Lu)

HOW SIMPLE IS AS SIMPLE AS POSSIBLE?

P. Heyes

Albert Einstein said "It can scarcely be denied that the supreme goal of all theory is to make the irreducible basic elements as simple and as few as possible without having to surrender the adequate representation of a single datum of experience". This statement is often paraphrased as "Everything should be made as simple as possible, but not simpler". This is a useful principle to apply to engineering models, though the engineer's aims may be more pragmatic than Einstein's. It can certainly be applied to the modelling of fatigue damage where the need is for answers accurate and reliable enough to support design decisions, but achievable at a reasonable cost and in a timely manner. This work considers the question in the title in connection with the characterization and modelling of fatigue performance in short glass fibre reinforced polyamides – materials widely used in automotive and other industries.

HBM United Kingdom Ltd (HBM Prenscia), Advanced Manufacturing Park Technology Centre, Brunel Way, Catcliffe, Rotherham, S60 5WG
Tel: 07785995244, Email: peter.heyes@hbmprenscia.com

A PROGRESSIVE DAMAGE FATIGUE MODEL FOR UNIDIRECTIONAL LAMINATED COMPOSITES BASED ON FINITE ELEMENT ANALYSIS: THEORY AND PRACTICE

M. Hack[1], D. Carrella-Payan[1], B. Magneville[1], T. Naito[2], Y. Urushiyama[2], T. Yokozeki[3], W. Yamazaki[3] & W. Van Paepegem[4]

The simulation of the fatigue damage of laminated composites under multi-axial and variable amplitude loadings has to deal with several new challenges and several methods of damage modelling. In this paper we present how to account for variable amplitude and multi-axial loading by using the damage hysteresis operator approach for fatigue. It is applied to a fatigue model for intra-laminar damage based on stiffness degradation laws from Van Paepegem and has been extended to deal with unidirectional carbon fibres. The parameter identification method is presented here and parameter sensitivities are discussed. The initial static damage of the material is accounted for by using the Ladevèze damage model and the permanent shear strain accumulation based on Van Paepegem's formulation. This approach has been implemented into commercial software (Siemens PLM). This intra-laminar fatigue damage model combines efficient methods with a low number of tests to identify the parameters of the stiffness degradation law, this overall procedure for fatigue life prediction is demonstrated to be cost efficient at industrial level. In an outlook we show, how to apply this model also to inter-laminar fatigue effects, which can lead to delamination.

[1]Siemens PLM Software, Kaiserslautern (Ger), Leuven (Be)
michael.hack@siemens.com
[2]Honda R&D Co Ltd, Tochigi
[3]Department of Aeronautics and Astronautics, The University of Tokyo
[4]Department of Materials Science and engineering, Ghent University

FATIGUE BEHAVIOUR OF NYLON-CLAY HYBRID NANOCOMPOSITES

S. J. Zhu*[1], A. Usuki[2] & M. Kato[2]

Hybrid organic-inorganic composites exhibit performance superior to those of their separate components, particularly if the inorganic components are dispersed in the organic matrix on a nanometer length scale. Clay is comprised of silicate layers having a 1 nm thick planar structure, which can be dispersed at the molecular level (nanometer level) in a polymer matrix with the polymer existing between the silicates layers. In 1987, nylon 6-clay hybrid (NCH) nanocomposites were synthesized in Toyota Central R & D Labs. The NCH nanocomposites showing high strength, elastic modulus, heat resistance and other properties have been used for automobiles, electronic industry and food package. It is expected that the applications of the nanocomposites will be extended to aerospace, energy, environment and biology industries if their time-dependent performances are evidenced. Therefore, cyclic deformation and fracture of NCH nanocomposites have been investigated.

Three kinds of materials used for fatigue tests were nylon 6, 2 wt% clay-reinforced nylon and 5 wt% clay-reinforced nylon nanocomposites. The specimens for tests are 80 mm in gage length, 10 mm in width and 4 mm in thickness. The fatigue tests were conducted at room and high temperatures (35 and 50 °C) using a servo-hydraulic mechanical testing machine. The fatigue tests with a load control were performed at stress ratio of 0.1 and frequency of 0.1 Hz in sine wave form. Fractured specimens were observed by scanning electron microscope and optical microscope. Cracks and fracture profiles on lateral surfaces were observed after being mounted in epoxy resin and polished with a doctor-lap. Electron Probe Micro Analyzer was used to analyze elemental distribution.

The fatigue strength of 2 wt% clay reinforced nylon is also increased by about 30% compared to that of nylon 6, but the fatigue strength of 5 wt% clay reinforced nylon is slightly increased or similar to that of nylon 6 at room temperature. The cyclic creep deformation is noted by examining using stress-strain hysteresis loops. The small cyclic deformation in 5 wt% clay reinforced nylon composite is attributed to the no increase in fatigue strength. The observation of fatigue fracture surfaces also shows brittle fracture features in 5 wt% composite, where crack initiated at small pores and a mirror zone can be seen. The fatigue fracture surfaces in 2 wt% composite show dimples and flowing morphology, implying a good compatibility of strength with ductility. Although the tensile strength and fatigue strength of NCH-2 are the highest at room temperature, the tensile strength and fatigue strength of NCH-5 became the highest at 35 °C. Fatigue fracture surfaces showed different patterns between at room temperature and 35 °C.

[1]Fukuoka Institute of Technology, Japan
[2]Toyota Central R & D Labs, Inc., Japan
*Corresponding author: zhu@fit.ac.jp

FATIGUE LIMITS OF NATURAL RUBBER USING CARBON BLACK FILLER MATERIALS

M. J. Jweeg[1], M. Al-Waily[2], D. C. Howard[3] & H. Y. Ahmad[4]

In this work, the effect of carbon black filler particles on the tensile and fatigue characteristics of natural rubber has been studied experimentally. The mechanical properties and fatigue characteristics of rubber were assessed with three different proportions (80, 90 and 100) of carbon black filler particles measured by Parts per Hundred by weight of Rubber (phr). The results show that the tensile and fatigue strength as well elongation are improved as the percentage of carbon black filler is increased. The tensile tests were carried out using the ASTM D412 standard test method at strain rate of 300 mm/min. The fatigue tests were carried out according to the ASTM D 813 (Standard Test Method for Rubber Deterioration-Crack Growth).

[1]Ministry of Higher Education & Scientific Research, Telafer University, Iraq. muhsin. e-mail: jweeg@uotelafer.edu.iq
[2]Ministry of Higher Education & Scientific Research, University of Kufa, Faculty of Engineering, Mechanical Engineering Department, Iraq. e-mail: muhanedl. alwaeli@uokufa.edu.iq
[3]Engineering Consultant, Mechanical Design, Safran Electrical & Power, Pitstone, Buckinghamshire, UK. e-mail: darren.howard@Sarfangroup.com
[4]Engineering Consultant, Stress & Materials, Safran Electrical & Power, Pitstone, Buckinghamshire, UK. e-mail: hayder.ahmad@Sarfangroup.com

15. DESIGN AND ASSESSMENT

TARGET RELIABILITY AS AN ACCEPTANCE CRITERION FOR FATIGUE

K. Wright

The assessment of a component's suitability for cyclic operation by demonstrating a fatigue usage factor of less than unity against an S-N fatigue design curve is the traditional and standard approach adopted by many industries. The ASME Section III boiler and pressure vessel code approach developed in the 1960s would appear to have served the nuclear industry well. However, the emerging understanding of environmental degradation on fatigue life and its significant dependence on temperature and strain rate for austenitic stainless steel has challenged the view held by many that the traditional approach being followed was 'fully' deterministic. The conservatism from the use of many assessment input variables and methodology assumptions is generally unquantified. Hence when combined with a fatigue design curve, now exacerbated by environmental penalty factors, this can result in analytical justification problems for new designs or when previous assessments are updated. This paper outlines the methodology developments and difficulties in the introduction of a more modern acceptance criterion for nuclear component fatigue assessments.

Chief Stress Engineer – Submarines, Rolls-Royce plc, PO Box2000, Derby, UK
Email: keith.wright@rolls-royce.com

MULTIAXIAL FATIGUE STRENGTH PREDICTION CRITERIA – DEVELOPMENT OF A VALIDATION DATA SET

J. Papuga*[1] , M. Lutovinov[1] & M. Růžička[1]

The paper describes various Internet data sources of experimental data related to fatigue. The goal of our work was to find a set of experimental data suitable for evaluating the predictive capability of various calculation methods for high-cycle multiaxial fatigue strength assessment. If it is sufficiently large and sufficiently demanding, this validation data set will provide a solid basis for checking the properties of existing calculation methods. The validation data set described by Papuga in [1], which is also available on-line in the FatLim database (www.pragtic.com/experiments. php), has been found to be burdened by errors due to inadequate checks on data quality.

None of the on-line systems described here offers either enough functionality or enough quality for the validation data set to be derived solely from it. The output of a search for adequate data items to build the data set is therefore described here. The criteria used for selecting the data items for the full set are defined, so that the same practice can be required in derived works. The final data set comprises 287 data items of fatigue strengths related to various multiaxial and uniaxial load combinations, including the mean stress effect. Only unnotched geometries of specimens are admitted to the data set, so that the focus of the validation will relate above all to the load multiaxiality effect, and the effect of the stress gradient is minimized.

Because a data set of such an extent is not easily transferable, and the individual items can be referred to only indirectly, a smaller validation data set has been prepared. Based on the assumption that only data items of adequate quality remained in the test set, the calculations were made using 22 different calculation methods (most of them are described in [1]). The relative errors of all methods in predicting the fatigue strengths when compared with the experimental inputs are statistically analysed for each item in the validation data set. The final small-scale validation data set marked as AMSD20 is proposed on the basis of the average value and standard deviation.

All 76 individual items in the final AMSD20 validation data set are provided in the paper. The paper also shows the statistical properties of the predictions of four calculation methods: Crossland, Liu-Zenner, Dang Van and Papuga PCr. The AMSD20 data set can therefore be used for designing and evaluating new calculation methods in the domain of multiaxial fatigue strength assessment. The practical importance of the data set is determined by the way in which it has been put together: (1) only primary sources of experimental data are used; (2) individual data items are assumed to be good enough (e.g. no more fatigue limits based on 1-3 data points, extrapolations not allowed, etc.) to the extent that the primary reports on the experiments are correct; (3) only data items with the largest average deviations from a perfect prediction or with the biggest scatter of prediction results remain; (4) the set is acceptably small - it should not take much time to analyse it, even when very complicated methods are used, or when only poor hardware or slow programmes are available.

[1]U12105 FME, Czech Technical University in Prague, Technická 4, 166 07 Prague 6, Czech Republic
*Corresponding author: Mr. Jan Papuga, +420 737 977 741, papuga@pragtic.com

APPROACH TO A GENERAL BENDING FATIGUE STRENGTH DESIGN METHOD

G. Deng[1] & T. Nakanishi[1]

In present bending fatigue strength design standards, the stress, so-called nominal stress, at the critical point where a fatigue crack initiates is calculated on the basis of material mechanics and modified using the stress concentration factor. Since material mechanics is only applicable to very simple geometries and the stress concentration factor can only be obtained for some typical geometries and load conditions, the nominal stress cannot be calculated for many machine components. In addition, the initiation and growth of a fatigue crack are influenced by not only the surface stress but also by the stress distribution depended by the geometry of the machine component, so, strictly speaking, the publicized bending fatigue strengths are only applicable to the machine components with the same geometry to the specimens used for estimating the fatigue strength. Consequently, the bending fatigue strength design cannot yet be performed for wide geometries of machine components that lead to many accidents so far.

The purpose of this study is to devise a bending fatigue strength evaluation method that can be widely used for machine component design, under the considerations that the bending fatigue breakage depends on the initiation of a surface fatigue crack and the conditions for the initiation of a fatigue crack can be used as the criteria for bending fatigue strength design. The specially designed and manufactured three bending specimens with crowned round notches are used in bending fatigue experiments. The obtained bending fatigue strengths are expressed as the maximum actual stress on the notch surface, which increase with the decrease in the notch radius. The stresses of the specimens were calculated with FEM, and the effects of the stress distributions on the fatigue strength were clarified. Considering the effects of the stress distribution on the bending fatigue strength, a method for estimating the actual tension fatigue strength of the smooth specimen is presented, which is the shape-independent fatigue strength and can be applicable to the bending fatigue strength design for the machine components with complex geometries.

[1]University of Miyazaki, Faculty of Engineering, Department of Mechanical Design Systems Engineering, 1-1 Gakuen Kibanadai-nishi, Miyazaki Shi, Miyazaki 889-2192, Japan t0d114u@cc.miyazaki-u.ac.jp